T0224204

Communications
in Computer and Information Science 650

Commenced Publication in 2007
Founding and Former Series Editors:
Alfredo Cuzzocrea, Dominik Ślęzak, and Xiaokang Yang

Editorial Board

More information about this series at http://www.springer.com/series/7899

Huajun Chen · Heng Ji
Le Sun · Haixun Wang
Tieyun Qian · Tong Ruan (Eds.)

Knowledge Graph and Semantic Computing

Semantic, Knowledge, and Linked Big Data

First China Conference, CCKS 2016
Beijing, China, September 19–22, 2016
Revised Selected Papers

 Springer

Editors
Huajun Chen
Zhejiang University
Zhejiang
China

Heng Ji
Rensselaer Polytechnic Institute
Troy, NY
USA

Le Sun
Chinese Academy of Sciences
Beijing
China

Haixun Wang
Google Research
Mountain View, CA
USA

Tieyun Qian
Wuhan University
Wuhan, Hubei
China

Tong Ruan
East China University of Science and
   Technology
Shanghai
China

ISSN 1865-0929   ISSN 1865-0937 (electronic)
Communications in Computer and Information Science
ISBN 978-981-10-3167-0   ISBN 978-981-10-3168-7   (eBook)
DOI 10.1007/978-981-10-3168-7

Library of Congress Control Number: 2016957858

Printed on acid-free paper

This Springer imprint is published by Springer Nature
The registered company is Springer Nature Singapore Pte Ltd.
The registered company address is: 152 Beach Road, #22-06/08 Gateway East, Singapore 189721, Singapore

Preface

This volume contains the papers presented at CCKS 2016: the China Conference on Knowledge Graph and Semantic Computing held during September 19–22, 2016 in Beijing.

CCKS is organized by the Technical Committee on Language and Knowledge Computing of CIPS (Chinese Information Processing Society). CCKS 2016 was the merger of two premier relevant forums held previously: the Chinese Knowledge Graph Symposium (KGS) and the Chinese Semantic Web and Web Science Conference (CSWS). KGS was first held in Beijing in 2013, in Nanjing in 2014, and in Yichang in 2015. CSWS was first held in Beijing in 2006, and has been the main forum for research on the Semantic (Web) technologies in China for nearly ten years. The new conference, CCKS, brings together researchers from both forums and covers wider fields including the knowledge graph, the Semantic Web, linked data, NLP, knowledge representation, graph databases etc. It aims to become the top forum on knowledge graph and semantic technologies for Chinese researchers and practitioners from academia, industry, and government.

The theme of this year was "Semantic, Knowledge and Linked Big Data."

There were 82 submissions. Each submission was reviewed by at least two, and on average 2.5, Program Committee members. The committee decided to accept 21 full papers and eight short papers. The program also included four invited keynotes, four tutorials, four shared tasks, one panel, and one industrial forum. The CCIS volume contains revised versions of 12 full papers, 7 short papers, and 6 shared tasks. This year's talks were given by Prof. Ian Horrocks from Oxford University, Prof. Gerhard Weikum from Max-Planck-Institut für Informatik, Dr. Haixun Wang from Facebook, and Prof. Heyan Huang from Beijing Institute of Technology. The tutorials were given by Dekang Lin from Sigularity.io, Jie Bao from MemeCT, Jeff Pan from Aberdeen University, Tong Ruan from East China University of Science and Technology, Haixun Wang from Facebook, Zhongyuan Wang from Microsoft Research Asia, and Wei Hu and Gong Cheng from Nanjing University.

The hard work and close collaboration of a number of people have contributed to the success of this conference. We would like to thank the members of the Organizing Committee and Program Committee for their support; and the authors and participants who are the primary reason for the success of this conference.

Finally, we would like to appreciate the sponsorships from TRS and Unisound as golden sponsors, and Baidu, Fujitsu, and Puhui Finance as silver sponsors.

August 2016

Le Sun
Haixun Wang
Huajun Chen
Heng Ji
Jiaoyan Zhu
Wei Hu
Haofen Wang
Jie Bao
Kang Liu
Zhichun Wang
Yuan Ni
Qi Zhang
Xianpei Han
Yiqun Liu
Jinguang Gu
Tieyun Qian
Tong Ruan
Honghan Wu
Xiangwen Liao

Organization

Conference General Chairs

Le Sun   Software Institute, China Academy of Science, China
Haixun Wang   Facebook, USA

Program Committee Chairs

Huajun Chen   Zhejiang University, China
Heng Ji   Rensselaer Polytechnic Institute, USA

Tutorial Chairs

Xiaoyan Zhu   Tsinghua University, China
Wei Hu   Nanjing University, China

Industry Forum Chairs

Haofen Wang   East China University of Science and Technology, China
Jie Bao   Memect, China

Evaluation Chairs

Kang Liu   Institute of Automation, China Academy of Science, China
Zhichun Wang   Beijing Normal University, China

Poster/Demo Chairs

Yuan Ni   IBM Research, China
Qi Zhang   Fudan University, China

Local Chairs

Xianpei Han   Software Institute, China Academy of Science, China
Yiqun Liu   Tsinghua University, China

Sponsorship Chairs

Jinguang Gu   Wuhan University of Science and Technology, China

Publication Chairs

Tieyun Qian	Wuhan University, China
Tong Ruan	East China University of Science and Technology, China

Publicity Chairs

Honghan Wu	King's College London, UK
Xiangwen Liao	Fuzhou University, China

Area Chairs

Gong Cheng	Nanjing University, China
Jianfeng Du	Guangdong University of Foreign Studies, China
Yansong Feng	Peking University, China
Yu Hong	Suzhou University, China
Zhiyuan Liu	Tsinghua University, China
Gerard de Melo	Tsinghua University, China
Yao Meng	Fujitsu, Japan
Jeff Pan	Aberdeen University, UK
Guilin Qi	Southeast University, China
Bin Qin	Harbin Institute of Technology, China
Xipeng Qiu	Fudan University, China
Quan Wang	China Academy of Science, China
Xin Wang	Tianjin University, China
Gang Wu	Northeastern University, China

Program Committee

Xipeng Qiu	Fudan University, China
Kun Xu	Peking University, China
Dezhao Song	Thomson Reuters, USA
Xiaowang Zhang	Tianjin University, China
Ran Yu	L3S, Germany
Jianfeng Du	Guangdong University of Foreign Studies, China
Yansong Feng	Peking University, China
Xin Wang	Tianjin University, China
Zhigang Wang	Tsinghua University, China
Jie Lu	IBM, USA
Yankai Lin	Tsinghua University, China
Quan Wang	Chinese Academy of Sciences, China
Gang Wu	Northeastern University, China
Songfeng Huang	IBM Research, China
Saisai Gong	Nanjing University, China
Chengjie Sun	Harbin Institute of Technology, China
Yu Hong	Soochow University, China

Contents

Linked Data and Knowledge-Based Systems

CCKS 2016 Shared Tasks

Knowledge Representation and Learning

A Joint Embedding Method for Entity Alignment of Knowledge Bases

Yanchao Hao$^{(\boxtimes)}$, Yuanzhe Zhang, Shizhu He, Kang Liu, and Jun Zhao

National Laboratory of Pattern Recognition, Institute of Automation,
Chinese Academy of Sciences, Beijing 100190, China
{yanchao.hao,yuanzhe.zhang,shizhu.he,kang.liu,jun.zhao}@nlpr.ia.ac.cn

Abstract. We propose a model which jointly learns the embeddings of multiple knowledge bases (KBs) in a uniform vector space to align entities in KBs. Instead of using content similarity based methods, we think the structure information of KBs is also important for KB alignment. When facing the cross-linguistic or different encoding situation, what we can leverage are only the structure information of two KBs. We utilize seed entity alignments whose embeddings are ensured the same in the joint learning process. We perform experiments on two datasets including a subset of Freebase comprising 15 thousand selected entities, and a dataset we construct from real-world large scale KBs – Freebase and DBpedia. The results show that the proposed approach which only utilize the structure information of KBs also works well.

Keywords: Embeddings · Multiple knowledge bases · Structure information · Freebase · DBpedia

1 Introduction

As the amount of knowledge bases (KBs) accumulated rapidly on the web, the problem of how to reuse these KBs has gained more and more attention. In the real-world scenarios, many KBs describe the same entities in different ways, because KBs are distributional heterogeneous resources created by different individuals or organizations. For example, president *Barack Hussein Obama* is denoted by *m.02mjmr* in Freebase [3], while *Barack_Obama* in DBpedia [2]. Aligning such same entities could help people acquire knowledge more conveniently, as they no longer need to look up multiple KBs to obtain the full information of an entity. However, knowledge base alignment is not a trivial task, and the alignment system is often complex [8,15]. Many traditional KB matching pipeline systems including [7,11,20,22] are based on content similarity calculation and propagation.

There are some standard benchmark datasets from the Ontology Alignment Evaluation Initiative (OAEI), on which several alignment systems perform alignment algorithms. The datasets don't contain many relationships and two KBs to be aligned have common relation and property strings, which can be used

© Springer Nature Singapore Pte Ltd. 2016
H. Chen et al. (Eds.): CCKS 2016, CCIS 650, pp. 3–14, 2016.
DOI: 10.1007/978-981-10-3168-7_1

to compute content similarity to assist instances alignment. The statistics of the author-disambiguation dataset from OAEI2015 Instance Matching are as Table 1. Think about a real case, we have an entity named *m.02mjmr* refering to president *Barack Hussein Obama*, How do we align it with the entity named *Barack_Obama* in another KB with all of the relations and properties in two different encoding system? When facing the cross-linguistic or different encoding situation, what we can leverage are only the structure information of two KBs. Content information is important to KB alignment, but we think the structure information of KBs is also significant. Based on the observation above, we create two datasets including a subset of Freebase comprising 15 thousand selected entities (FB15K) and a dataset we construct from real-world large scale KBs: Freebase and DBpedia. What we try to do is to construct datasets with abundant relations and rich structure information, regardless of the content.

Table 1. Statistics of author-dis sandbox from OAEI2015. The relations and properties are shared in two KBs.

Instance class	Author-instance	Relation	Property
2	854	6	6

In this paper, we perform the KB entity alignment task by leveraging the embeddings of the KBs which are learned via the structure of KBs no matter what the content is. In previous work, KB embeddings [4–6,9,17,21] are learned in order to complete the KB, and they aim at single KB. If the embedding learning method is applied on two KBs, we will obtain two independent embeddings in two different vector spaces. To represent two KBs in a uniform embedding vector space, we give some initial alignments, called seed entity alignments. In the learning process, we ensure the embeddings of the seed entities try to maintain the same. In this way, we could jointly learn the embeddings of the two KBs in a uniform embedding vector space, with two KBs connected by the seed entities "bridge". The seed alignments help learn potential alignments of the two KBs in the uniform expressive vector space via the network of the triplets. Entities with similar learned embeddings could be considered as the same entities. Thus we could find more alignments. The proposed method does not depend on manually designed rules and features, and we do not need to be aware of the content of the KBs. As a result, the proposed approach is more adaptive, could be easily utilized to large scale applications.

We conduct extensive large scale experiments on two datasets including a subset of Freebase comprising 15 thousand selected entities, denoted FB15K [5], and a dataset we construct from real-world large scale KBs – Freebase and DBpedia. The results indicate that the proposed method could achieve promising performance, and the joint embedding method only utilize the structure information of KBs, which may be a efficient supplement for KB alignment pipeline systems.

To the best of our knowledge, this is the first work to deal with the KB alignment problem using an end to end joint embedding model only utilizing the structure information of KBs. In summary, the contributions of this paper are as follows.

(1) We propose a novel model which jointly learns the embeddings of multiple KBs in a uniform vector space to align entities in KBs, only using the structure information of KBs.
(2) We construct two datasets for KB alignment task based on real-world large scale KBs: FB15K datasets and DBpedia-Freebase datasets, which have abundant relationships and rich structure information.
(3) We conduct experiments on the datasets, and the experimental results show that our approach works well.

The remainder of this paper is organized as follows. We first introduce our task in detail and overview of the related work. Then, we present the proposed method in the following section. Finally, we show the experimental results and conclude this paper.

2   Background

2.1   Task Description

Entity alignment on KBs, which is to align the entities that referring to the same real-world things, has been a hot research topic in recent years. For example, we should align the entity *m.02mjmr* in Freebase with the entity *Barack_Obama* in DBpedia. The goal of the KB alignment is to link multiple KBs effectively and create a large scale and unified KB from the top-level to enrich the KBs, which can be used to help machines understand the data and build more intelligent applications.

KBs usually use Resource Description Framework Schema (RDFS) or Ontology Web Language (OWL) or triples to describe ontology, defining elements such as "class", "relation", "property", "instance" and so on. The research of KB alignment starts from ontology matching [23–25], mainly focusing on the semantic similarity at early time.

2.2   Related Work

Over the years, various methods have been proposed for KB alignment. Akbari et al. [1] and Suna et al. [19] utilize string-matching based methods which are quite straightforward but fail when two entity mentions are crossing languages or significantly different in literal. Joslyn et al. [10] consider the aligning problem as a graph homomorphism problem, [14,16] exploit Instance-based techniques to align KBs, and some take the KB alignment as combinatorial optimization problems [13].

In pairs-wise alignment methods, some supervised learning methods compare vectors via property to judge an entity pair whether should be aligned or not.

This kind of technology contains decision tree [26], Support Vector Machine (SVM) [27], ensemble learning [28] and so on. Some clustering based methods [29] learns how to cluster similar entities better.

In collective alignment methods, [18] present a PARIS system based on probabilistic method to align KBs without tuning parameters and training data, but PARIS cannot handle structural heterogeneity. Lacoste et al. [12] propose SiGMa algorithm to propagate similarity via viewing the task of KB alignment as a greedy optimization problem of global match score objective function.

All of them are based on content similarity calculation and propagation, and many ontology matching pipeline systems including [7, 11, 20, 22] which participate in the OAEI 2015 Instance Matching track need to calculate content similarity. Some of them use local structure information to propagate similarity, but from another point of view, we think that the global structure information of KBs is also important. Our proposed models are based on global structure information of KBs, regardless of what the content exactly is.

3   Datasets

Because of the lack of suitable data for our task which is under the cross-linguistic or different encoding situation, we construct two datasets based on real-world large scale datasets. Firstly we present a dataset generated from FB15K, which is extracted from Freebase comprising 15 thousand selected entities. Then we illustrate the DBpedia-Freebase dataset (DB-FB), which are extracted from DBpedia and Freebase.

3.1   FB15K Dataset

FB15K. There are more than 2.4 billion triplets and 80 million entities in Freebase[1]. The base dataset we choose should not be too small to acquire enough overlapping part, and should not be too large to cause computational bottlenecks. As a tradeoff, we choose FB15K containing 592,213 triplets with 14,951 entities and 1,345 relationships. We randomly split them into two KBs, i.e., $kb1$ and $kb2$, with a large amount of overlapping part. Given a ratio number, i.e., the parameter $splitRatio$, we split the intersecting entities into two parts. The first part remains identical entity mention forms in two KBs, denoted as remaining part (seed alignment part). The second part keeps the entity mention forms unchanged in $kb1$, and changes the entity mention forms in $kb2$ by suffixing a certain string like "_#NEW#" to create the different entities, denoted by changing part (target alignment part), which is used for evaluation. Figure 1 indicates the splitting process of our datasets. There are two advantages of our proposed dataset. First, since they origin from the same FB15K dataset, we can control the overlapping part conveniently. Second, the gold entity alignment is known, so the evaluation is more accurate.

---

[1] https://developers.google.com/freebase/.

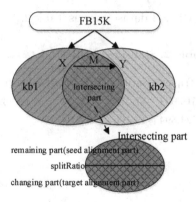

Fig. 1. The process of splitting FB15K.

3.2   DB-FB Dataset

DB-FB. There are more than 3 billion factual triples in DBpedia[2] and 2.4 billion in Freebase. DBpedia also provide datasets which contain triples linking DBpedia to many other datasets. Based on the given entity alignments with Freebase released on the DBpedia website[3], we can build a DBpedia-Freebase alignment dataset. Following the original intention, we intend to construct a dataset with abundant relationships and rich structure information. The dataset we construct should not be too small to contain enough structure information, and too large to cause computational bottlenecks. The steps of constructing DB-FB dataset are as follows.

Step1. As we know, Freebase triples have some Compound Value Types (CVTs) to represent data where each entry consists of multiple fields. Firstly, we need to convert the triples in Freebase which contain CVT to factual triples by reducing the CVT in the preprocessing step.

Step2. Then we find the triples in DBpedia and Freebase whose head and tail entity both show up in the given alignments.

Step3. In the selected triples, we count the frequencies of the entity alignment pairs (take the Napierian logarithm of the product of each entity's frequency in a pair) and rank the frequencies of the entity pairs.

Step4. Based on the top 10 thousand most frequently showing up entity alignment pairs, we select the triples whose head entity or tail entity are among the top 10 thousand entity alignment pairs in the picked out triples in step2.

Step5. Then we make a filter to reduce the triples whose entity frequency are less than 7 in DBpedia and 35 in Freebase.[4]

---

[2] http://wiki.dbpedia.org/Downloads2015-10.

[3] http://downloads.dbpedia.org/2015-10/links/freebase_links.nt.bz2.

[4] In step5, 7 and 35 are empirical values chosen in experiments.

The statistics of the DB-FB dataset are as Table 2.

Table 2. Statistics of DB-FB dataset.

	Triples	Entities	Relations	Align_pairs
DB	515,937	57,076	373	13,932
FB	724,894	19,166	1,219	

4 Methodology

Given two KBs, denoted by $kb1$ and $kb2$ respectively. The facts in both KBs are represented by triplets (h, r, t), where $h \in E$ (the set of entities) is the head entity, $t \in E$ is the tail entity, and $r \in R$ (the set of relationships) is the relationship. For example, *(Obama, president_of, USA)* is a fact. Different from previous KB embedding learning methods, our model learns the joint embeddings of the entities and the relations of two KBs. In detail, we firstly generate several entity alignments using simple strategies which leverage some extra information or other measures. As shown in Fig. 2, the entities in the same color are the entity alignments, i.e., the selected seed entities. In this way, the seed entity alignments could serve as bridges between $kb1$ and $kb2$, thus we can learn the joint embeddings of both KBs in a uniform framework.

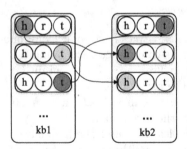

Fig. 2. Selecting seed entities in two KBs.

A KB is embedded into a low-dimensional continuous vector space while certain properties of it are preserved. Generally, each entity is represented as a point in that space while each relation is interpreted as an operation over entity embeddings. For instance, TransE [5] interprets a relation as a translation from the head entity to the tail entity. Following the energy-based framework in TransE, the energy of a triplet is equal to $d(h+r, t)$ for some dissimilarity measure d, which we take to be either the L_1 or L_2-norm. To learn such embeddings, we minimize the margin-based objective function over the training set:

$$L = \sum_{(h,r,t)\in S} \sum_{(h',r,t')\in S'_{(h,r,t)}} \{[\gamma + d(h+r,t) - d(h'+r,t')]_+ +$$

$$\lambda_1 \sum_{y\in\{h,h',r,t,t'\}} |\|y\|_2 - 1|\} + \lambda_2 \sum_{(e_i,e'_i)\in A} \|e_i - e'_i\|_2 \tag{1}$$

where $[x]_+$ denotes the positive part of x, $\gamma > 0$ is a margin hyper-parameter, λ_1, λ_2 are ratio hyper-parameters, A is the selected seed alignments whose entities are represented by e_i in $kb1$ and e'_i in $kb2$, and

$$S'_{(h,r,t)} = \{(h',r,t)|h' \in E\} \cup \{(h,r,t')|t' \in E\} \tag{2}$$

The set of corrupted triplets, constructed according to Eq. (2), is composed of training triplets with either the head or tail replaced by a random entity (but not both at the same time). The objective function is optimized by stochastic gradient descent (SGD) with mini-batch strategy. The soft constraints of the entities and relations (the λ_1 part in Eq. (1)) is important because they are meaningful in preventing the training process to trivially minimize the loss function by increasing the embedding norms and shaping the embeddings [5]. The alignment part (the λ_2 part in Eq. (1)) helps learn the alignment information between KBs.

Following the projection transformation idea, we can fix Eq. (1) by adding a projection transformation matrix M_d:

$$L = \sum_{(h,r,t)\in S} \sum_{(h',r,t')\in S'_{(h,r,t)}} \{[\gamma + d(h+r,t) - d(h'+r,t')]_+ +$$

$$\lambda_1 \sum_{y\in\{h,h',r,t,t'\}} |\|y\|_2 - 1|\} + \lambda_2 \sum_{(e_i,e'_i)\in A} \|M_d e_i - e'_i\|_2 \tag{3}$$

The projection matrix M_d serves as the transformation of different KB vector spaces. It is more reasonable to transfer one KB vector space to another when we want to connect two KBs.

In the learning process, the embeddings of the entities in $kb1$ could become more and more similar with the same factual world entities in $kb2$ through seed entities. So the jointly learned embeddings can help improve entity alignment between the two KBs. The key of our model is to align two KBs using embeddings in a uniform space that jointly learned via the overlapping parts between the two KBs.

5   Experimental Evaluations

5.1   Baseline

Given the two KBs generated from FB15K, we suffix all the intersecting entities in $kb2$ to make $kb2$ totally different from $kb1$. Then we learn the embeddings of the entities and relations in the two KBs in two vector space individually following TransE [5]. Since the intersecting entities are split into two parts, we use the remaining part to learn the projection transformation matrix M, representing transformation of the same entities from one vector space to the other using the following equations:

$$Y^T = MX^T \tag{4}$$

$$M = Y^T X (X^T X)^{-1} \tag{5}$$

Where X denotes the embedding matrix of the remaining part of $kb1$, Y denotes the embedding matrix of the remaining part of $kb2$, and M denotes the projection transformation matrix. Let len denote the number of entities in the remaining part, and dim denotes the dimension of the embeddings. So the matrixes X and Y are $\mathbb{R}^{len \times dim}$, while the matrix M is $\mathbb{R}^{dim \times dim}$.

As for the changing part, we could obtain the projection embeddings of the entities of $kb1$ Y in the vector space of $kb2$, using Eq. (4). In other words, the function of matrix M is to transform the embeddings in $kb1$'s vector space to $kb2$'s vector space in order to find the degree of similarity between the projected embeddings and the true embeddings.

In DB-FB dataset, we can directly use the Eqs. (4) and (5) without changing the forms of the entities.

5.2 Implementation

For our model, we regard the remaining part as the seed alignment part. Some hyper-parameters in two models were just set empirically. For experiments settings, when we learn the embeddings, we choose the margin γ as 1, the dimension k as 100, the λ_1 in loss function as 0.1, the λ_2 in loss function as 1, the epoch for training as 2000. The dissimilarity measure d is L_2 distance. The embeddings of entities and relations are initialized in the range of $[-0.01, 0.01]$ with uniform distribution. Table 3 shows the comparison of overall results where there are 7,365 entities in the target entity part for evaluation and 14,825 entities in $kb2$ totally under the parameters setting $splitRatio = 0.5$. Every entity in the target entity part could have rank value from 1 to 14,285. In this table, Mean_Rank represents the mean rank value of the target entities part, and Hits@n means the ratio number of entities that rank at top n.

Table 3. Overall results of FB15K. JE denotes our joint Embedding model in Eq. (1), and $JEwP$ denotes as our joint Embedding model with projection matrix in Eq. (3).

Models	Mean_Rank	Hits@1	Hits@10	Hits@100
Baseline	95.97	23.96%	54.96%	83.22%
JE	94.76	29.73%	56.36%	81.91%
JEwP	**88.51**	**29.88%**	**59.21%**	**84.97%**

Our model improves the performance significantly compared with the baseline approach. We believe that the good performance of our model is due to jointly embedding two KBs into a uniform vector space via seed entities "bridge" connecting two KBs. The seed alignments help learn potential alignments of the two KBs in the uniform expressive vector space via the triplets' network, while in the baseline model, we can only utilize the projection transformation matrix learned from the seed alignment part with no extended alignment information on the whole.

Table 4. Effect of *splitRatio* on FB15K.

Models	splitRatio	Mean_Rank	Hits@1	Hits@10	Hits@100
Baseline	0.1	91.79	25.10%	56.52%	83.84%
	0.3	92.71	23.34%	54.25%	82.95%
	0.5	95.97	23.96%	54.96%	83.22%
	0.7	94.44	25.12%	55.66%	83.10%
JE	0.1	352.00	10.25%	20.19%	47.18%
	0.3	239.56	15.47%	31.63%	63.30%
	0.5	94.76	29.11%	56.62%	81.91%
	0.7	97.85	29.73%	56.36%	81.48%
JEwP	0.1	205.74	17.59%	42.34%	66.67%
	0.3	123.28	25.63%	55.35%	78.60%
	0.5	88.51	29.88%	59.21%	84.97%
	0.7	**86.83**	**30.38%**	**60.70%**	**85.14%**

We also explore the effect of *splitRatio*, i.e., the number of seed entities, on our models. As shown in Table 4, along with the ascending order of *splitRatio*, the Mean_Rank value of our model decreases and the Hits@n increases, indicating the performance of our model getting better because of more seed entities. While the baseline model shows much more placid when the *splitRatio* increases, as shown in Fig. 3. The impression of the baseline model is that the performance should be increasing along with the ascending order of *splitRatio* because there are more and more data to learn the projection transformation matrix M well. But the result is almost placid. The reason in further analysis shows that when *splitRatio* = 0.1 the categories of the entities in the remaining part to learn are already covered enough and the projection transformation based method cannot depict the influence of different relations to the entity alignment. While our joint embedding method learns the different representations of different relations which help improve the performance of alignment. For example, the relation "son_of" is more important than the relation "nationality" in judging whether two entities are the same or not.

We conduct experiments on the DB-FB dataset, and the results are as Table 5. The baseline model has better $Mean\_Rank$, and our joint embedding

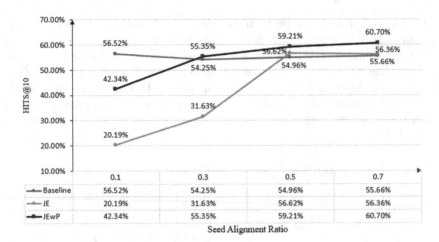

Fig. 3. The performance of our models on FB15K along with the ascending *splitRatio*.

Table 5. Results on the DB-FB dataset.

Models	SeedAlignments_Ratio	Mean_Rank	Hits@1	Hits@10	Hits@100
Baseline	0.1	554.43	2.20%	14.56%	45.81%
	0.3	485.46	2.00%	14.85%	46.46%
	0.5	490.23	2.18%	14.76%	48.11%
	0.7	**476.25**	2.20%	14.81%	49.13%
JE	0.1	1019.98	0.66%	4.46%	27.06%
	0.3	785.63	1.25%	8.55%	35.67%
	0.5	723.18	1.57%	11.35%	38.68%
	0.7	700.00	1.87%	13.15%	41.40%
JEwP	0.1	639.89	1.91%	9.86%	41.72%
	0.3	605.39	2.38%	15.18%	45.65%
	0.5	524.74	3.91%	19.39%	53.34%
	0.7	510.18	**4.64%**	**19.90%**	**54.89%**

projection model has better performance at $Hits@n$ when we have a certain number of seed Alignments. The reason may be that the baseline model learns the projection transformation matrix from a global perspective, while our models learn the embeddings of KBs and projection matrix M_d (especially the JEwP model) in the iterative optimization process. The DB-FB dataset is relatively large and the selected DBpedia set which has 515,937 triples and 57,076 entities is more sparse than the selected Freebase set which has 724,894 triples and 19,166 entities. So on the DB-FB dataset, it may be more difficult to capture the global accurate alignment information for our models in the learning process. Note that our models only utilize the structure information of KBs to align entities, not the accurate content information. When we are faced with actual KB alignment task, our model may be an efficient supplement to the alignment pipeline systems.

6 Conclusions

We propose a model which jointly learns the embeddings of KBs in a uniform vector space via seed entity alignments to align KBs. Generally, our model with projection matrix has better performance than our model without projection matrix, which is reasonable for that projection matrix indicates transformation of KBs, and projection matrix should be added when we associate one vector space with another. To utilize structure information of KBs, we construct two datasets including FB15K and DB-FB based on real-world large scale KB. The experimental results show that the proposed approach which only utilize the structure information of KBs also works well, and may be an efficient supplement for KB alignment pipeline systems.

Acknowledgement. This work was supported by the Natural Science Foundation of China (No. 61533018), the National Basic Research Program of China (No. 2014CB340503) and the National Natural Science Foundation of China (No. 61272332). And this work was also supported by Google through focused research awards program.

References

1. Akbari, I., Fathian, M., Badie, K.: An improved mlma+ and its application in ontology matching. In: Innovative Technologies in Intelligent Systems, Industrial Applications, CITISIA 2009, pp. 56–60. IEEE (2009)
2. Auer, S., Bizer, C., Kobilarov, G., Lehmann, J., Cyganiak, R., Ives, Z.: DBpedia: a nucleus for a web of open data. In: Aberer, K., Choi, K.-S., Noy, N., Allemang, D., Lee, K.-I., Nixon, L., Golbeck, J., Mika, P., Maynard, D., Mizoguchi, R., Schreiber, G., Cudré-Mauroux, P. (eds.) ASWC/ISWC -2007. LNCS, vol. 4825, pp. 722–735. Springer, Heidelberg (2007). doi:10.1007/978-3-540-76298-0_52
3. Bollacker, K., Evans, C., Paritosh, P., Sturge, T., Taylor, J.: Freebase: a collaboratively created graph database for structuring human knowledge. In: Proceedings of the ACM SIGMOD International Conference on Management of Data, pp. 1247–1250. ACM (2008)
4. Bordes, A., Weston, J., Collobert, R., Bengio, Y.: Learning structured embeddings of knowledge bases. In: Conference on Artificial Intelligence, number EPFL-CONF-192344 (2011)
5. Bordes, A., Usunier, N., Garcia-Duran, A., Weston, J., Yakhnenko, O.: Translating embeddings for modeling multi-relational data. In: Advances in Neural Information Processing Systems, pp. 2787–2795 (2013)
6. Chang, K.-W., Yih, W.-T., Meek, C.: Multi-relational latent semantic analysis. In: EMNLP, pp. 1602–1612 (2013)
7. Damak, S., Souid, H., Kachroudi, M., Zghal, S.: EXONA Results for OAEI (2015)
8. Gokhale, C., Das, S., Doan, A., Naughton, J.F., Rampalli, N., Shavlik, J., Zhu, X.: Hands-off crowdsourcing for entity matching. In: Proceedings of the ACM SIGMOD International Conference on Management of Data, pp. 601–612. ACM (2014)

9. Ji, G., He, S., Xu, L., Liu, K., Zhao, J.: Knowledge graph embedding via dynamic mapping matrix. In: Proceedings of ACL, pp. 687–696 (2015)
10. Joslyn, C.A., Paulson, P., White, A., Al Saffar, S.: Measuring the structural preservation of semantic hierarchy alignments. In: Proceedings of the 4th International Workshop on Ontology Matching, CEUR Workshop Proceedings, vol. 551, pp. 61–72. Citeseer (2009)
11. Khiat, A., Benaissa, M.: InsMT+ Results for OAEI 2015 Instance Matching (2015)
12. Lacoste-Julien, S., Palla, K., Davies, A., Kasneci, G., Graepel, T., Ghahramani, Z.: Simple greedy matching for aligning large knowledge bases. In: Proceedings of the 19th ACM SIGKDD International Conference on Knowledge Discovery and Data Mining, pp. 572–580. ACM (2013)
13. Prytkova, N., Weikum, G., Spaniol, M.: Aligning multi-cultural knowledge taxonomies by combinatorial optimization. In: Proceedings of the 24th International Conference on World Wide Web Companion, pp. 93–94. International World Wide Web Conferences Steering Committee (2015)
14. Pushpakumar, R., Sai Baba, M., Madurai Meenachi, N., Balasubramanian, P.: Instance Based Matching System for Nuclear Ontologies (2016)
15. Scharffe, F., Zamazal, O., Fensel, D.: Ontology alignment design patterns. Knowl. Inf. Syst. **40**(1), 1–28 (2014)
16. Seddiqui, M., Nath, R.P.D., Aono, M.: An efficient metric of automatic weight generation for properties in instance matching technique. arXiv preprint arXiv:1502.03556 (2015)
17. Socher, R., Chen, D., Manning, C.D., Ng, A.: Reasoning with neural tensor networks for knowledge base completion. In: Advances in Neural Information Processing Systems, pp. 926–934 (2013)
18. Suchanek, F.M., Abiteboul, S., Senellart, P.: Paris: probabilistic alignment of relations, instances, and schema. Proc. VLDB Endow. **5**(3), 157–168 (2011)
19. Suna, Y., Maa, L., Wangb, S.: A Comparative Evaluation of String Similarity Metrics for Ontology Alignment (2015)
20. Wang, W., Wang, P.: Lily Results for OAEI 2015 (2015)
21. Wang, Z., Zhang, J., Feng, J., Chen, Z.: Knowledge graph embedding by translating on hyperplanes. In: AAAI, pp. 1112–1119. Citeseer (2014)
22. Zhang, Y., Li, J.: RiMOM Results for OAEI 2015 (2015)
23. Shvaiko, P., Euzenat, J.: Ten challenges for ontology matching. In: Meersman, R., Tari, Z. (eds.) OTM 2008. LNCS, vol. 5332, pp. 1164–1182. Springer, Heidelberg (2008). doi:10.1007/978-3-540-88873-4_18
24. Berstein, P.A., Madhavan, J., Rahm, E.: Generic schema matching, ten years later. Proc. VLDB Endow. **4**(11), 695–701 (2011)
25. Shvaiko, P., Euzenat, J.: Ontology matching: state of the art and future challenges. IEEE Trans. Know. Data Eng. **25**(1), 158–176 (2013)
26. Han, J.W., Kambe, M.: Data Mining: Concepts and Techniques. Morgan Kaufmann, San Francisco (2006)
27. Vapnik, V.: The Nature of Statistical Learning Theory. Springer, Berlin (2000)
28. Kantardzic, M.: Data Mining. Wiley, Hoboken (2011)
29. Cohen, W.W., Richman, J.: Learning to match, cluster large high-dimensional data sets for data integration. In: Procedings of Advances in Neural Information Processing Systems, pp. 905–912. MIT Press, Cambridge (2005)

A Multi-dimension Weighted Graph-Based Path Planning with Avoiding Hotspots

Shuo Jiang[1,3(✉)], Zhiyong Feng[1,3], Xiaowang Zhang[2,3], Xin Wang[2,3], and Guozheng Rao[2,3]

[1] School of Computer Software, Tianjin University, Tianjin 300350, China
{jiangshuo,zyfeng}@tju.edu.cn
[2] School of Computer Science and Technology,
Tianjin University, Tianjin 300350, China
{xiaowangzhang,wangx,rgz}@tju.edu.cn
[3] Tianjin Key Laboratory of Cognitive Computing and Application,
Tianjin 300350, China

Abstract. With the development of industrialization rapidly, vehicles have become an important part of people's life. However, transportation system is becoming more and more complicated. The core problem of the complicated transportation system is how to avoid hotspots. In this paper, we present a graph model based on a multi-dimension weighted graph for path planning with avoiding hotspots. Firstly, we extend one-dimension weighted graphs to multi-dimension weighted graphs where multi-dimension weights are used to characterize more features of transportation. Secondly, we develop a framework equipped with many aggregate functions for transforming multi-dimension weighted graphs into one-dimension weighted graphs in order to converse the path planning of multi-dimension weighted graphs into the shortest path problem of one-dimension weighted graphs. Finally, we implement our proposed framework and evaluate our system in some necessary practical examples. The experiment shows that our approach can provide "optimal" paths under the consideration of avoiding hotspots.

Keywords: Path planning · Avoiding hotspots · Multi-dimension weighted graph · Shortest path problem

1 Introduction

1.1 Path Planning

Path planning is a sequential algorithm based on existing nodes, edges and weights according to a certain method. These nodes, edges and weights are data in graph model, which can represent different things in different situations, such as obstacles, hotspots and so on. Path planning technology has been applied extensively in many domains since it was proposed [1]. There are plenty of applications in frontier domains: route planning of unmanned aerial vehicles, robot path planning and space path planning of rocket launch. This technology not only speeds up the progress in frontier domains, but also becomes an integral part of our daily life [12]. For example, GPS navigation helps

© Springer Nature Singapore Pte Ltd. 2016
H. Chen et al. (Eds.): CCKS 2016, CCIS 650, pp. 15–26, 2016.
DOI: 10.1007/978-981-10-3168-7_2

us plan path while we are driving. The application of the technology in business and management domain is logistics, that is, resources dispatch in a reasonable way. Generally speaking, the problems which can be translated into graph models can be translated into nodes, edges and weights, and we can use path planning to solve them [2].

1.2 Avoiding Hotspots

Avoiding hotspots is a way to use existing data to deal with hotspots, thus making the overall planning immune to the effects of hotspots. Avoiding hotspots is not eliminating hotspots. What we avoid is the effect and damage caused by hotspots. That is to reduce the occurrence probability of hotspots. Therefore, avoiding hotspots can be used in path planning, especially vehicle routing problem.

Vehicle routing problem (VRP) was proposed firstly in 1959. It means a distribution center which provides different numbers of cargoes to a certain number of customers in a city or an area. The most important part is to plan an optimal path, the goal of which is to reach the highest economic benefit under the precondition that the requirements of customers must be met. There are some requirements in path planning such as the shortest path, the shortest time or the least oil consumption.

The loss is not only in economy, but also in environment. Fuel consumption causes air pollution. Traffic jam happens frequently in our daily life. The probability of traffic accidents is still rising. Thus, it is of great importance to avoid hotspots.

1.3 Related Works

Since the weighted graph was proposed, plenty of applications have been generated. After researching the current situation of path query of weighted graph, we reach some conclusions as follows: [3] proposes a new optimal route search model for public transit based on directed weighted graph. This model cannot only allow users to set their ideal maximum walking distance, but also meet the requirements of personalized query by using the flexible weighting graph method, especially the strong expression ability for multi-object query. Fire brigades are always required to reach the field of fire in the shortest time. Therefore, selecting the appropriate path can effectively reduce the loss of casualties and property. [4] establishes a model of multi-stage weighted directed graph aimed at this problem. Multi-stage weighted directed graph is a common graph, which can translate a lot of practical matters, such as transportation, engineering, and management, into the shortest path problem. [5] establishes a general weighted graph for transportation. It is a math model which combines network analysis and linear programming theory. This model solves the practical matter caused by network complexity, path diversity and load capacity.

There is a query of weighted graph based on regular expression in [6]. The author characterizes the query of weighted graph and proposes the algorithm. This query can be embedded effectively in XML query language. [7] proposes a query of weighted regular expression. This query can allow users to define priority of weight and be connected naturally with link information of quantitative database. The authors also propose a distribution algorithm to calculate this query. This algorithm can also solve the multi-source shortest path problem in case that we do not know the complete graph. In order to query and analyze graph database by the method of aggregate function and order, [8] extends a previous graph query language. This language can support query probability graph in that way. [13] presents a SPARQL-based querying language named pSPARQL for probabilistic RDF graphs.

We can see that there is a preliminary study on weighted graph from the research above. Not only the language is of normalization and flexibility, but also the algorithm for the weighted graph is of efficiency. However, there is still something that can be extended in the weighted graph to make it more effective than previous ones. We can see that the previous studies on weighted graph only focus on one-dimension weight, while the traffic environment is complex in the real world, which means one-dimension weight cannot describe the information exactly. As we know, many cases are composed of factors influenced with each other. Therefore, it is unreasonable to calculate weighted graph with one-dimension weight.

We can also see from the research on related works that the weighted graph systems can not meet all the requirements of customers. The problems we face every day are all in characteristics for ourselves, as a result, the previous one-dimension weighted graph models can only solve a little part of the problem. Therefore, there is a lack of model or an aggregate function which users can define for their own demands to solve problems.

1.4   Overview

The overview of the paper is as follows: we focus on complex traffic environment with multi-dimension weighted graph; then we establish a model according to the specific circumstances and requirements; and we define an aggregate function which can translate multi-dimension weighted graph into one-dimension weighted graph; finally we use Dijkstra algorithm, a classical path planning algorithm, to solve the problem and propose a good plan.

The overall structure of this paper is as follows: Sect. 1 mainly introduces the related work, the lack of them, and how we deal with the lack through our innovations of this paper; Sect. 2 introduces related concept of graph; Sect. 3 introduces the multi-dimension weighted graph and the aggregate function; Sect. 4 simulates a specific case, then we show how to use our method to solve it; Sect. 5 shows the whole framework, related experience and the efficiency of this framework; Sect. 6 concludes this paper and the future work.

2   Graph

2.1   Basic Definitions

Graph is a math object to describe the relationship among objects. Assume graph G is an ordered two tuple (V, E), and V represents a set of vertices, then we can use $V(G)$ to represent a set of nodes; E represents a set of edges. Similarly, we can use $E(G)$ to represent a set of edges. Note that E and V do not intersect. The elements of E are all two tuple, which are noted by (x, y), and $x, y \in V$ [9].

Path is a sequence from one node to another. For example, assume that a path P is $v_0, (e_1, v_1, e_2, v_2, \ldots, e_k, v_k)$ and the length of this path is k. There is a pair (v_{i-1}, v_i), which is an edge from v_{i-1} to v_i. If the starting node and the ending node is the same, then we say this path is close. Otherwise, we say it is open.

Graphic model is a structure model, whose function is to describe a system. Constituted by nodes and edges, it can represent everything in the real world, so it can be used to describe the relationship among all objects. Therefore, a graphic model is a good tool for modeling, and it proposes a good way to deal with complex systems.

2.2   Directed Graph

Directed graph is a subclass of graph. Every edge is directed in directed graph. Directed graph is an ordered pair. Assume there is a directed graph D, and the ordered pair is (V, E). Then V is a nonempty set constituted by nodes of D. The elements in V are vertexes. E is a set of edges of D constituted by V.

Every element in the edge of directed graph is an ordered pair. Assume that an ordered pair is $<u, v>$ in directed graph D, which we say is a directed edge. u represents the starting node of the edge, while v represents the ending node of the edge. Therefore, $<u_i, v_i>$ and $<v_i, u_i>$ represent two different edges.

2.3   Undirected Graph

Undirected graph is a subclass of graph, however, different from directed graph. Every edge in undirected graph is undirected, and it is represented by unordered pair. Assume that an undirected graph $G = <V, E>$. V is a nonempty set constituted by nodes. E is a set of unordered two tuple constituted by the elements in V, and it is a set of edges. Intuitively, if all edges in a graph are undirected, then the graph is undirected. Unordered pair is usually noted by round brackets. Contrary to directed graph, there are no starting node s and ending node s in undirected graph. That is, the two unordered pairs (v_i, v_j) and (v_j, v_i) present the same edge.

2.4   Weighted Graph

Weighted graph is also a subclass of graph, but it is different from the previous two graphs for the reason that every edge in weighted graph is assigned with a value. This value is the weight of this edge. Weight can take a certain value to represent other objects, such as cost, probability and so on. Broadly speaking, weight in the weighted graph is usually single.

Assume there is a weighted graph $G = <V, E, W>$. V is a nonempty set constituted by nodes, then it is a set of nodes of G. E is a set of two tuple constituted by the elements in V, then it is a set of edges. W represents weight. If E is constituted by a set of unordered two tuples, then the weighted graph G is an undirected weighted graph.

Otherwise, it is directed weighted graph. The study of this paper focuses on undirected weighted graph.

3 Extension of Weighted Graph

We can see that there is usually a single weight in weighted graph from the previous research. However, objects are usually affected by more than one factor in the real world. For example, when a user needs a path to reach the destination in the shortest time, we should consider about the length, the probability of traffic jam and the degree of traffic jam. Not only that, different people will have different requirements for the problem of path planning. Some people need the shortest time to reach the destination, while some people need the shortest length to reach there. Therefore, faced with many different requirements, we cannot use the single weight to solve those problems, but need to define a multi-dimension weighted graph to create different models to solve different practical problems.

3.1 Multi-dimension Weighted Graph

Multi-dimension weighted graph is an extension of weighted graph. From the Sect. 2 we have already known that weighted graph G can be represented as follows:

$$G = <V, E, W> \tag{1}$$

Multi-dimension weighted graph is not a single weight on every edge. Assume a graph G_1 is a multi-dimension weighted graph. Then it can be represented as follows:

$$G_1 = <V, E, (w_1, w_2, \ldots)> \tag{2}$$

Every weight in multi-dimension weighted graph is related to path planning, since the study is based on path planning. Here shows an example based on the G_1.

Assume that G_1 represents a graph of probability of traffic jam. Then V represents a set of location in a city; E, represents a set of roads; $w_1, w_2, \ldots$ represents a set of attributes of the roads. In other words, they are the factors which can affect the path planning. There are three weights w_1, w_2 and w_3, where w_1 represents the length of every road; w_2 represents the degree of traffic jam; w_3 represents the probability of traffic jam of every road.

Therefore, we can connect the related factors to solve the problem in the real world more exactly. Then we will introduce the aggregate function $f(x)$ to deal with these weights.

3.2 Aggregate Function

Aggregate function $f(x)$ can calculate several weights, and obtain the functional results. That is, we can use aggregate functions to translate several weights into one weight.

There are several common aggregate functions in Excel, such as addition, subtraction, multiplication, division and averaging. For example, the addition aggregate function:

$$f(w_1, w_2, \ldots) = w_1 + w_2 + \cdots \tag{3}$$

The common aggregate functions are too restricted, which can only calculate common data. Users usually face difficult situations, and these aggregate functions can not deal with them well. Therefore, we need to propose aggregate functions for users to calculate special problems for their own demands. If we have the aggregate function which is defined by ourselves, then we can translate the multi-dimension weighted graph $G_1 = <V, E, (w_1, w_2, \ldots)>$ into one-dimension weighted graph $G = <V, E, w_f>$ by the aggregate function $f(w_1, w_2, \ldots)$, and $w_f = f(w_1, w_2, \ldots)$.

4   Application of Multi-dimension Weighted Graph

We focus on a problem of navigation based on a graph of probability of traffic jam to introduce the two concepts proposed in the third chapter in detail.

4.1   Graph of Probability of Traffic Jam

We establish a graph of probability of traffic jam in order to solve the problem of path planning for those people who are in emergency. This model can reduce the risk to meet the traffic jam. This model not only has the basic information of roads and locations, but also has the attributes which will affect the traffic jam for every road. Graph of probability of traffic jam is a multi-dimension weighted graph. We define it as follows:

$$G = <V, E, (w_l, w_h, w_p)> \tag{4}$$

- V represents the locations in the city. We note $A, B, C, \ldots$ to represent them.
- E represents the roads in the city. We note $a, b, c, \ldots$ to represent them.
- (w_l, w_h, w_p) represents three-dimension weights, where w_l represents length of road. We note L, and $L \in (0, +\infty)$; w_h represents the degree of traffic jam. We note H, and $H = \{1, 2, 3\}$ (1 represents a weak degree, 2 represents a common degree, 3 represents a strong degree); w_p represents the probability of traffic jam. We note P, and $P \in [0,1]$.

We study the case of traffic jam in the real world, then we define the following aggregate function $f(x)$:

$$f(w_l, w_h, w_p) = w_l \times (w_h \times w_p + 1) \tag{5}$$

Then we use the above function to calculate the three-dimension weights to obtain the result. We will establish a graph model to show how to obtain this result.

4.2   Data and Results

We establish 5 nodes and 6 edges. The detail data and the graph model are as follows:

- $V = \{A, B, C, D, E\}$,
- $E = \{a, b, c, d, e\}$,
- The set of three-dimension weights of the 6 edges is $W = \{(12, 1, 0.1), (11, 2, 0.5),$ $(1, 3, 0.8), (6, 2, 0.3), (15, 2, 0.2), (5, 2, 0.6)\}$

Figure 1 shows the graph model which stores the above data.

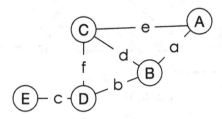

Fig. 1. The graph model

First, put W into the aggregate function $f(x)$, which we have defined before.

For example, $w_a = (12, 1, 0.1)$, then according to the $f(w_l, w_h, w_p) = w_l \times (w_h \times w_p + 1)$, $w_{fa} = 12 \times (1 \times 0.1 + 1) = 13.2$. Then we deal with the result by rounding to get the integer 13. After calculating the three-dimension weights by aggregate function, we get the final $w_f = \{13, 22, 3, 9, 21, 11\}$. Finally we calculate the final result w_f with Dijkstra algorithm [10] to get the final value from every node to other nodes. Table 1 shows the case from node A to other nodes.

Table 1. Result of Dijkstra (A) of graph of probability of traffic jam

B	C	D	E
13	21	32	35

From the aggregate function we can see that the bigger the value is, the higher risk to meet the traffic jam will be.

5   Experiments

5.1   Framework

We show the whole framework in our architecture [11]. First, according to the special problems and different requirements, we establish the suitable models with the related factors, which will affect the result in the real world. Then, consider the relationship among these weights to define the aggregate function $f(w_1, w_2, \ldots)$. After that, put the

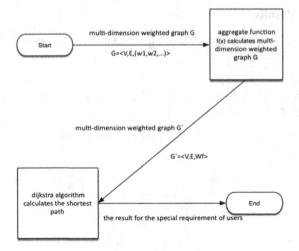

Fig. 2. The framework

weights of multi-dimension weighted graph in the aggregate function to get the final result of weight. This process can realize the translation from multi-dimension weighted graph to one-dimension weighted graphs. Finally, we use the Dijkstra algorithm to get the result which we need. Figure 2 shows the framework.

We will put some more examples to show that the model can solve a lot of practical problems.

5.2   Efficiency

According to the graph of probability of traffic jam, we test the efficiency with the following number data size 10, 20, 30, 40 and 50 and time the corresponding time of program running. Then Fig. 3 shows the result of efficiency.

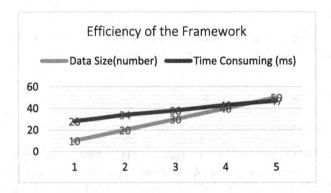

Fig. 3. The efficiency of the framework

From the result we can see that the slope of rising data size is bigger than the slope of rising time consuming. Therefore, the efficiency of the model plays an important role in the age of big data.

5.3 Graph of Traffic Accidents

We want to choose a safer path rather than the shortest path when we go to a dangerous place. According to the traffic accidents, we define the following model:

$$G = \ <V, E, (w_l, w_q, w_v, w_k)> \tag{6}$$

- V represents the locations in the city. We use A, B, C, ...to represent them.
- E represents the roads in the city. We note a, b, c, ... to represent them.
- (w_l, w_q, w_v, w_k) represents four-dimension weights, where w_l represents length of roads, and we note $L \in (0, +\infty)$; w_q represents the traffic volume. The more the volume is, the higher risk of accidents will be. We note $q = \{1, 2, 3\}$ (1 represents a small volume, 2 represents a middle volume, 3 represents a large volume); w_v represents the maximum speed (the faster the speed is, the more dangerous it will be), and we note $V \in (0, +\infty)$; w_k represents the risk factor, and we note $K \in (0,1)$.

According to the research of accidents, we define the following aggregate function to deal with the four weights:

$$f(w_l, w_q, w_v, w_k) = \frac{w_l * w_q}{100} * \frac{w_v}{(1 - w_k)} \tag{7}$$

We still use the earlier graph to make the experiment. We establish the following data:

- $V = \{A, B, C, D, E\}$,
- $E = \{a, b, c, d, e\}$,

And the set of four-dimension weights of the 6 edges is W = {(12, 1, 80, 0.1), (11, 2, 70, 0.5), (1, 3, 90, 0.8), (6, 2, 60, 0.3), (15, 2, 75, 0.2), (5, 2, 80, 0.6)}. First, put W into the aggregate function $f(x)$. After calculating the four-dimension weights by aggregate function, we get the final w_f = {10, 30, 13, 10, 28, 20}. Finally we calculate the final result w_f with Dijkstra algorithm to get the final value from every node to other nodes. Table 2 shows the value of risk from node A to other nodes.

Table 2. The result of Dijkstra (A) of graph of traffic accidents

B	C	D	E
10	20	40	53

According to the aggregate function we can see that the bigger the value is, the higher risk to meet the traffic accidents will be.

5.4   Graph of Traffic Cost

According to the framework, we establish a model for the user who care about the traffic cost. we define the following model:

$$G = <V, E, (w_l, w_x, w_e)>$$ (8)

(w_l, w_t, w_e) is three-dimension weight, where w_l represents length of roads, and we note $L \in (0, +\infty)$; w_t represents cost of consumed fuel of per kilometer, and we use $X \in (0, +\infty)$; w_v represents the maximum speed (the faster speed, the more dangerous), and we use $E \in (0, +\infty)$;

According to the research of cost, we define the following aggregate function to deal with the three weights:

$$f(w_l, w_t, w_e) = w_l \times w_x + w_e$$ (9)

We still use the earlier graph to make the experience. We establish the following data:

- $V = \{A, B, C, D, E\}$,
- $E = \{a, b, c, d, e\}$,
- The set of three-dimension weight of the 6 edges is $W = \{(15, 4, 20), (11, 2, 25), (8, 3, 15), (33, 2, 28), (15, 2, 22), (40, 2, 30)\}$.

First, put W into the aggregate function $f(x)$. After calculating the three-dimension weight by aggregate function, we get the final $w_f = \{80, 47, 39, 94, 52, 110\}$. Finally we calculate the final result w_f with Dijkstra algorithm to get the final value from every node to other nodes. Table 3 shows the value of cost from node A to other nodes.

Table 3. The result of Dijkstra (A) of graph of traffic cost

B	C	D	E
80	52	127	166

According to the aggregate function we can see that the bigger the value is, the higher cost spending on the path will be.

5.5   Graph of Traffic Time

According to the framework, we establish a model for the user who care about the traffic time. We define the following model:

$$G = <V, E, (w_l, w_v, w_d)>$$ (10)

(w_l, w_v, w_d) is three-dimension weight, where w_l represents length of roads, and we note $L \in (0, +\infty)$; w_v represents cost of consumed fuel of per kilometer, and we note $V \in (0, +\infty)$; w_d represents the value of traffic jam. As we know, the traffic time is related to the case of traffic jam. Therefore, we will use the previous result in this model. We note $D = \{13, 22, 3, 9, 21, 11\}$;

According to the research of cost, we define the following aggregate function to deal with the three weights:

$$f(w_l, w_v, w_d) = \frac{w_l}{w_v} * \frac{w_d}{(w_d - 1)} \tag{11}$$

We still use the previous graph to make the experience. We establish the following data:

- $V = \{A, B, C, D, E\}$,
- $E = \{a, b, c, d, e\}$,
- The set of three-dimension weights of the 6 edges is $W = \{(120, 40, 13), (110, 80, 22), (100, 20, 3), (60, 15, 9), (150, 70, 21), (50, 10, 11)\}$.

First, we put W into the aggregate function $f(x)$. After calculating the three-dimension weights by aggregate function, we get the final $w_f = \{3, 1, 7, 4, 2, 5\}$. Finally we calculate the final result w_f with Dijkstra algorithm to get the final value from every node to other nodes. Table 4 shows the value of time from node A to other nodes.

Table 4. The result of Dijkstra (A) of graph of

B	C	D	E
3	2	4	11

According to the aggregate function we can see that the bigger the value is, the higher time spending on the path will be.

6 Conclusions

Path planning is a problem which we are always researching and probing. Although the applications of path planning are emerging in an endless stream, there is no suitable application for general users for their personal requirements. This paper establishes a multi-dimension weighted graph to exactly realize the simulation of the practical problems in the real world. We put the factors which will affect each other together to constitute the multi-dimension weighted graph, then according to the relationship among the weights to define aggregate function, which can calculate the factors to meet the different requirements of different users.

This paper establishes a framework by the combination of multi-dimension weighted graphs and aggregate functions. Then we simulate a graph of probability of traffic jam to show the process of this framework. We improve the previous related works based on weighted graph with only one-dimension weight and imperfect aggregate functions. Finally, we make it more suitable to solve the problem of path planning in the real world.

We put some other examples such as the graph of traffic accidents, the graph of traffic cost and the graph of traffic time. Intuitively, the framework can solve a lot of problems, and it can regard the previous result as a factor in this model. We also define the corresponding aggregate function to calculate the three examples above.

We will improve the second processing module, in which we will use another algorithm to deal with the final weight in our future work. We hope the framework can solve a lot of practical problems beyond the path planning.

Acknowledgements. We would like to thank Yaqi Chen for previous survey and useful comments. This work is supported by the program of the National Key Research and Development Program of China (2016YFB1000603) and the National Natural Science Foundation of China (NSFC) (61502336, 61373035). Xiaowang Zhang is supported by Tianjin Thousand Young Talents Program.

References

1. Stiles, P., Glickstein, I.: Route planning. In: IEEE, pp. 420–425 (1991)
2. Zhang, G., Hu, X., Chai, J., Zhao, L., Yu, T.: Summary of path planning algorithm and its application. Mod. Mach. **5**, 85–90 (2011)
3. Yao, C., Li, X., Shen, L.: Weighted directed graph model for searching optimal travel routes by public transport. Appl. Res. Comput. **30**(4), 1058–1063 (2013)
4. Hao, R.: Fire rescue based on shortest route model and its solution strategies. China Sci. Technol. Inf. **19**, 29–30 (2010)
5. Feng, M.: The transportation problem based on general weighted graph. Math. Pract. Theory **38**(9), 131–135 (2008)
6. Flesca, S., Furfaro, F., Greco, S.: Weighted path queries on semistructured databases. Inf. Comput. **204**(5), 679–696 (2006)
7. Stefanescu, D., Thomo, A.: Enhanced regular path queries on semistructured databases. In: Grust, T., et al. (eds.) EDBT 2006. LNCS, vol. 4254, pp. 700–711. Springer, Heidelberg (2006). doi:10.1007/11896548_53
8. Dries, A., Nijssen, S.: Analyzing graph databases by aggregate queries. In: MLG 2010, pp. 37–45, July 2010 (2012)
9. Diestel, R.: Graph Theory, 4th edn. Tsinghua University Press, Beijing (2013)
10. Cormen, T.H., Leiserson, C.E.: Introduction to Algorithm, 2nd edn. Machinery Industry Press, Cambridge (2006)
11. Hellings, J., Kuijpers, B., Van den Bussche, J., Zhang, X.: Walk logic as a framework for path query languages on graph databases. In: Proceedings of ICDT 2013, Genoa, Italy, pp. 117–128. ACM (2013)
12. Zhang, X., Van den Bussche, J.: On the power of SPARQL in expressing navigational queries. Comput. J. **58**(11), 2841–2851 (2015)
13. Fang, H., Zhang, X.: pSPARQL: a querying language for probabilistic RDF (extended abstract). In: Proceedings of ISWC Posters and Demos 2016, Kobe, Japan (2016)

Position Paper: The Unreliability of Language - A Common Issue for Knowledge Engineering and Buddhism

Zhangquan Zhou$^{(\boxtimes)}$ and Guilin Qi

School of Computer Science and Engineering, Southeast University, Nanjing, China
{quanzz,gqi}@seu.edu.cn

Abstract. According to the studies of Kurt Gödel and Ludwig Wittgenstein, both of formal languages and human languages are unreliable. This finding inherently influences the development of artificial intelligence and knowledge engineering. On the other hand, their finding, i.e., the unreliability of languages, was early discussed by Gautama Buddha who founded Buddhism. In this paper, we discuss the issue of the unreliability of language by bridging the perspectives of Gödel, Wittgenstein and Gautama. Based on the discussion, we further give some philosophical thoughts from the perspective of knowledge engineering.

Keywords: Knowledge engineering · Artificial intelligence · Language · Unreliability · Gödel · Wittgenstein · Buddhism

1 Introduction

The core of *knowledge engineering* is to apply different kinds of formal languages (or models) to represent and manage human languages (or knowledge) [6]. Researchers in the filed of knowledge engineering develop and optimize methods for automatic knowledge management, and even for making knowledge machine-understandable, which is also one target of *artificial intelligence*. A question then arises naturally: *"Is it possible that all kinds of human knowledge can be represented and handled by machines?"* Unfortunately, the answer turns out to be "NO" from a theoretical perspective. The reason can be traced back to Gödel's incompleteness theorems [3], which state that there is not a complete and reliable system for proving all mathematical consequences. This result was further extended to general formal languages by Tarski [7]. Since all current methods of representing and handling human knowledge are based on formal languages or models, they submit to Gödel's incompleteness theorems.

From a philosophical perspective, even human languages are unreliable, i.e., they are full of contradictions and mistakes. This was claimed by Ludwig Wittgenstein, whose theories essentially laid the foundations of the *linguistic philosophy*. The findings of Gödel and Wittgenstein inherently influence the development of artificial intelligence and knowledge engineering. However,

© Springer Nature Singapore Pte Ltd. 2016
H. Chen et al. (Eds.): CCKS 2016, CCIS 650, pp. 27–32, 2016.
DOI: 10.1007/978-981-10-3168-7_3

their finding, i.e., the unreliability of languages, was early discussed by Gautama Buddha who founded Buddhism. Gautama said that we cannot rely on languages to understand the truth of the world. In summary, researchers in the filed of knowledge engineering have to face an issue: *(formal and human) languages are unreliable.*

The aim of this paper is not to address the above issue, i.e., the unreliability of languages, but to highlight this issue by bridging the perspectives of Gödel, Wittgenstein and Gautama. Based on the discussion of their views, we give some philosophical thoughts from the perspective of knowledge engineering.

2   The Incompleteness of Mathematical Languages

In 1931, Kurt Gödel published his incompleteness theorems (known as Gödel's incompleteness theorems) [3], which are important in both of the mathematical logic and the philosophy mathematics. As shown in the name of Gödel's incompleteness theorems, the famous theorems indicate an important property of mathematical languages: *incompleteness.* That is, one cannot prove all mathematical consequences by using the axioms expressed in mathematical languages[1]. We give a simple case of incompleteness in the following example.

Example 1. Consider a mathematical system δ where set *is the unique atomic element represented by capital letters. It is allowed that a set can be a member of other sets. The symbols "$:=, \{, |, \}, \notin$" and "$\in$" are also allowed in δ. These symbols have standard mathematical semantics and can help us to describe the relations between sets ($\notin$ or $\in$) and define new sets. We now define a set X by $X := \{S | S \notin S\}$ where S is a set.*

The mathematical system δ is such simple that it contains sets as its unique elements. However, δ is incomplete. Consider the question *"Is X a member of itself?"* (formally $X \in X$?). If $X \in X$, then it contradicts the definition given in Example 1. Thus X should not be in X; if $X \notin X$, then X satisfies the above definition and should be in X. From the standard semantics of $\in$ and $\notin$, any two sets should have a binary relation of either $\in$ or $\notin$. However, both of the two results ($X \in X$ and $X \notin X$) would result in contradictions. The problem in Example 1 is known as *Russell's paradox* (or more popular, *Barber paradox*) that was proposed by Bertrand Russell in 1903. One intuitive cause of this kind of problems is *self-reference* [1], i.e., X is also defined by itself.

Russell's paradox indirectly resulted in *the Third Mathematical Crisis* when the completeness of mathematical languages was under suspicion. In fact, to find a complete and perfect mathematical system was the dream of several famous mathematicians, like David Hilbert, who devoted a large part of his life to this work. However, Kurt Gödel finally broke the dream with a simple fact: mathematical languages are unreliable due to incompleteness.

---

[1] Alfred Tarski extended the results to more general formal system five years later. We refer the readers to Tarski's undefinability theorem [7].

3   Language-Game

Ten years before the publication of the incompleteness theorems, a young man, in his doctoral thesis [8], claimed that *even human languages are unreliable.*

The young man is Ludwig Wittgenstein, who was the protege of Bertrand Russell, and the classmate of Adolf Hitler. In his doctoral thesis [8], Wittgenstein analyzed the *contradictions, vagueness* and *woven* of human languages. The finding of Wittgenstein really resulted in the crisis of philosophy, and built a new branch of philosophy: the *linguistic philosophy.* The importance of linguistic philosophy lays in that, it fundamentally queries all the other schools of western philosophy. This is because that, all philosophical theories are described by human languages. Since human languages are unreliable and full of contradictions, philosophical theories cannot hold even from the level of language. To understand this, we give the following example.

Example 2. A Martian asked Wittgenstein a question: "Sir, how many toes do philosophers have?" Wittgenstein answered: "Of course ten!" The Martian raised his feet of only six toes and said sadly: "Does that mean, we Martians cannot be philosophers?"

In daily life, we use a large amount of *concepts* (or *terms*) in our languages to communicate with each other, like the concept *"philosopher"* in the above example. However, we rarely doubt the exact meaning of these used concepts (more precisely, their *intensions* and *extensions*). This is due to two reasons: (1) We tend to easily believe what we see and hear, which would further be reflected in our languages; (2) The meanings of concepts in our languages are established by people who live in the same environment around us. Recall Example 2. In our common sense, a philosopher should first be a human. Thus, we undoubtedly treat *"having ten toes"* as one of the extensions of the concept "philosopher".

Wittgenstein argued that, philosophical theories cannot be built on the concepts without exact definitions and specifications. For example, when one asks the question *"Who am I? Why am I in this world?"*, he should first give the exact definitions of the concepts *"I"*, *"world"* and the semantics of the interrogatives, *"who"* and *"why"*. He also found that, many concepts were even defined by themselves (this is like the case of self-reference mentioned in Sect. 2). In this sense, using languages is similar to playing games: given a bunch of words without meanings, we first set the rules of how to use these words, and then we use these words to communicate, to describe our ideas and react to the word usage of other people. This is also called *language-game* by Wittgenstein [8]. From his theories, language-game is such a process when the meanings of words are not static, but are dynamically changed according to different situations and different people. He also said that language-game is being played in every family where children learn to use languages from their parents.

The idea of language-game gives a negative signal: we can never find the truth of the world and ourselves through languages, since we are just being in a game where the rules of how to use languages are full of mistakes and contradictions. This finding brought Wittgenstein a huge suffering at the end of his life.

4   The Influence of Gödel's Incompleteness Theorems to Computer Science and Artificial Intelligence

Different from the suffering of Wittgenstein, Gödel's incompleteness theorems actually benefited scientists a lot. The incompleteness theorems and the contributions of many mathematicians for solving the Third Mathematical Crisis virtually gave birth to the strongest tool in human history, *computer*.

At the same time of Gödel, there was an American mathematician called Alonzo Church. He and Gödel contributed a lot to *recursion theory*. Driven by the similar dream of Hilbert, they started their journey to a different destination: to build a universal machine that can describe and solve mathematical problems. However, a Ph.D student of Church was the first one to reach the destination. The name of this student is Alan Turing. The universal machine described by Turing is also well known as *Universal Turing machine* [2], which is supposed to be the prototype of computer.

The prototype of computer further encouraged scientists to investigate whether a computer can solve all problems with termination. By referring to Gödel's incompleteness theorems, scientists immediately found the answer: "NO". It is proved that there exists a large group of problems that cannot be solved with termination [2]. These problems are also called *uncomputable problems*. The uncomputability can also be ascribed to the issue of self-reference (see the related content in Sect. 2). Many uncomputable problems are proved using the technique called *diagonalization*[2], which is essentially a formulation of self-reference.

Gödel's incompleteness theorems also influenced the development of artificial intelligence (AI for short). On one hand, the *symbolic logic* as the classical approach of AI suffers that, computers cannot solve some problems when using highly expressive logic languages with termination. With regard to techniques, the notion of self-reference is always used to identify the completeness of a logic language, e.g., introducing *canonical models* to identify completeness for *model theoretic logic*. On the other hand, many researchers pay more attention on *statistical models* rather than *symbolic logics*. The basic idea of statistical techniques is to make machines behave like men by *leaning* human behavior. The related techniques are known as *statistical learning* or *machine learning*. However, for both of symbolic logics and statistical models, researchers just choose to weaken the influence of the unreliability of formal languages or models, but not completely solve it.

5   Everything with Form is Unreal

From the previous sections, we can conclude that, both of formal languages and human languages are unreliable in some sense. Further, the strongest tool, computer, is not as such strong as we imagine.

---

[2] The details of diagonalization can be found at [2].

During the time (the latter half of the 20th Century) when many AI researchers turned to statistic methods, western philosophers also found a fact that many ideas in the linguistic philosophy were early discussed in Buddhism [4]. Gautama Buddha, who built Buddhism, said in different Buddhist sutras that language is unreliable and is really an obstruction for us to the Enlightenment.

According to the opinions of Gautama, we human beings begin to understand the world by mapping different meanings to what we see and hear. These meanings, also called *forms* by Buddhists, are always our subjective thoughts which are incomplete, full of contradictions, and cannot reflect the reality of the world. However we always tend to believe that such forms are real. For example, our ancestors believed that the earth was in the center of the universe for a long time, since the sunrise and sunset looked like that the sun was just moving around the earth (similar to the moon). Here, *"the earth was in the center of the universe"* is such a form that our ancestors mapped to what they saw.

It is obvious that human language is also a kind of form. We map different concepts (or words) to what we see and hear. As time passes, we tend to rely on such concepts to understand this world. However, Gautama said that we cannot define *truth* and *reality* using forms, since forms are just reflections of our mind. Further, Gautama gave a strong claim that *everything with form is unreal*. This claim is deduced in *The Diamond Sutra* and cited in other sutras.

Gautama underlined to his proteges that, if someone believes that there exists the Enlightenment, he will never reach the Enlightenment. In other words, "Enlightenment" is just a word consisting of 13 letters. Gautama also argued that there is even no *"self"*, i.e., "self" is just an illusioned concept in our mind. Gautama said that almost all kinds of *sufferings* came from our persistence in "self". However, "self" is just a concept, but not a real existence according to Gautama. Thus, in many practices of Buddhists, e.g., *meditation*, people train themselves to jump beyond the bound of language, the constraint of "self", and all the other forms.

Backing to the unreliability of language, it seems that we have not any progress on the question *"Who am I? Why am I in this world?"*. However, we indeed have a deeper understanding of this question and our languages. That is, our languages are unreliable.

6 Discussion

Due to the generation of data by sensor networks, social media and different organizations, there is an exponential growth of structured or semi-structured data [5]. In this background, the techniques of AI and knowledge engineering are being widely used to represent and manage data (or knowledge) for different domains. On the other hand, the issue of the unreliability of formal languages and human languages has to be faced as well. In this part, we try to give some philosophical thoughts from the perspective of knowledge engineering.

First, it is not appropriate to find the exact meaning of "intelligence". From the perspective of the linguistic philosophy, there does not even exist an exact and static definition of "intelligence". According to the idea of Gautama, "intelligence" may just be a word created by human and turns out to be a wishful thinking of human, rather than a nature existence. In this sense, it is not appropriate to use any formal language or model to explain what is "intelligence".

Second, we should trade off between completeness and incompleteness. Completeness is an important property to show whether the utilized formal languages and models are reliable. However, *incompleteness* is inevitable in the sense that we utilize highly expressive formal languages. There has been work where researchers carefully sacrifice the completeness (reliability) of the utilized logic languages to achieve a better computational efficiency for logic reasoning. The related method is also called *incomplete reasoning* or *approximate reasoning*.

Third, we should combine different forms for representing knowledge. According to the arguments of Gautama, any kind of form is unilateral, subjective, and a partial reflection of our mind. Thus, it is rewardless to rely on any form to understand this world and ourselves. However, we have to use languages, or different forms to represent and manage knowledge, and to communicate with other people. Therefore, it is better for us to combine different forms, i.e., different formal languages or models to represent and manage knowledge, rather than to be constrained in only one formal language or representation of knowledge.

7   Conclusions

In this paper, we briefly discussed the findings of Gödel and Wittgenstein. That is, both of formal languages and human languages are unreliable. We further strengthened this claim by introducing some views of Gautama. We finally discuss this issue from the perspective of knowledge engineering.

References

1. Bartlett, S.J.: Reflexivity: A Source Book in Self-reference. North-Holland/Elsevier Science Publishers, Amsterdam (1992)
2. Davis, M.D., Weyuker, E.J.: Computability, Complexity, and Languages - Fundamentals of Theoretical Computer Science. Computer Science and Applied Mathematics. Academic Press, Cambridge (1983)
3. Gödel, K.: Über formal unentscheidbare sätze der principia mathematica und verwandter systeme i. J. Monatsh. Math. Phys. **38**, 173–198 (1931)
4. Gudmunsen, C.: Wittgenstein and buddhism. J. Int. Assoc. Buddh. Stud. **3**, 122–126 (1980)
5. Kleiner, N., Sejdovic, S., Zander, S., Setzer, T., Studer, R., Jähnichen, S.: Big data, smart data and semantic technologies (BDSDST). In: Proceedings of GI-Jahrestagung, pp. 1169–1170 (2015)
6. Studer, R., Benjamins, V.R., Fensel, D.: Knowledge engineering: principles and methods. J. Data Knowl. Eng. **25**(1–2), 161–197 (1998)
7. Tarski, A., Woodger, J.: The concept of truth in formalized languages. J. Corcoran **8**, 153–278 (1931)
8. Wittgenstein, L.: Tractatus Logico-Philosophicus, vol. 7. Routledge and Kegan Paul, London (1922)

Construction of Domain Ontology
for Engineering Equipment Maintenance
Support

YongHua Zeng[✉], JianDong Zhuang, and ZhengLian Su

College of Field Engineering, PLA University Science and Technology,
Nanjing 210000, China
Protege_user@126.com, zhuangjiandong1991@163.com,
182247847@qq.com

Abstract. According to the problem in the domain of engineering equipment maintenance, such as more knowledge points, broad scope, complex relationships, difficult in sharing and reuse, this paper put forward the category and professional field of engineering equipment maintain ontology, and analyzed knowledge source, extracted eight core concepts such as case, product, function, damage, environment, phenomena, disposal and resource, and formed concept hierarchy model further, and then analyzed data properties and object properties of core concepts, and tried to construct the engineering equipment maintain ontology with protege4.3, which put a solid foundation for the knowledge base and engineering equipment maintenance application ontology.

Keywords: Domain ontology · Maintenance support · Engineering equipment

1 Introduction

With the rapid development of engineering equipment and its maintenance support information construction, the degree of informatization improved continually, the maintenance support knowledge source on engineering equipment increased rapidly. In order to share and reuse the knowledge from different kind and different structure information system, and to meet the requirement of integration of joint security, it is urgent to strengthen management of engineering equipment maintenance knowledge.

Engineering equipment maintenance knowledge involves many disciplines such as mechanical engineering, electrical engineering, cybernetics, behavioral science and diagnosis technology. And there are many store kinds such as audio, video, model, animation, document, table and application software or system, but it has not unified description way, which will lead maintenance personnel to feel it is hard to find the related resources rapidly and precisely, and which also will lead Engineering equipment maintenance knowledge will not be able to applied effectively [1].

Ontology provides the clear, formal and specification explain of shared concept model, which can explain the semantic in an explicit and formal way. Ontology can improve the interoperability of high different structure system, which will lead to knowledge be shared and reused efficiently. So, the construction of engineering

© Springer Nature Singapore Pte Ltd. 2016
H. Chen et al. (Eds.): CCKS 2016, CCIS 650, pp. 33–38, 2016.
DOI: 10.1007/978-981-10-3168-7_4

equipment maintenance ontology will be benefit of sharing and reusing engineering equipment maintenance knowledge.

Ontology can be divided into top ontology, domain ontology, mission ontology and application ontology. Domain ontology is a professional ontology for special science, which definite the concepts and relationships of concepts, and describe the basic principles, main entities and activity relationships. Domain ontology provides the public understanding foundation, which is thought to be the most promising method to solve the information and knowledge island. Ontology is the concept basis and meta-model of knowledge base. In order to build an engineering equipment maintenance knowledge system successfully, this paper try to build an engineering equipment maintenance ontology preliminary closely in combination with the demand of engineering equipment repair.

2   Overview of Domain Ontology Construction

2.1   Principle

In the long practices of ontology construction, people have advanced many principles. The most influential principle was put forward by Tom Gruber in 1995, which concludes: clarity and objectivity, consistency, extensibility, minimum coding preferences and minimum ontology commitments [2]. Engineering equipment maintenance ontology construction will obtain this principle.

2.2   Tools and Methods

Domain ontology construction is an very onerous and complex system engineering. There are more than 60 building tools, but there is not a standard method. And domain ontology can't be automatically built, which can only be built by special peoples. We select protege4.3 as building tool and select seven steps as built method.

3   Construction of Equipment Maintenance Domain Ontology

Fully use for reference the soul of ontology construct methodology, we try to construct engineering equipment maintenance support domain ontology, combined with the circular iterative idea of circular obtain methodology, with the step of "seven steps", and with the method of engineering item management on 'spire prototype method'. Material steps as follows.

3.1   Nail Down Professional Category and Domain

As we know, engineering equipment maintenance pays most attention on engineering equipment's damage of using phase and related products, situation and repair. The ontology's user main concludes equipment maintenance support personnel, designer, developer, users and teaching staff in colleges and universities training institutions. The aim of engineering equipment maintenance ontology is to organize the maintenance knowledge with ontology idea and description language, which provides the realization of the knowledge representation [3].

3.2   Comb the Resources of Domain Knowledge

Ontology consists of five elements: concepts, relations, functions, axiom and examples. Concepts can form a classification level, can express the relationship, and can constraint through the relations, functions and axiom. According to the elements of ontology, we get the basic knowledge resources of engineering equipment repair through analysis.

(1) First resource: authoritative dictionary and encyclopedia. For example, we can get the definition of engineering equipment maintenance, and we can get the definition of equipment damage, equipment maintain from 《military language》 .

(2) Second resource: related domain thesaurus. For example, we can get the concept classification system and knowledge hierarchy relationship such as parts.

(3) Third resource: domain experts. In the view of some unclear concepts and relations, we can ask engineering equipment repair domain experts for confirm.

(4) Fourth resource: standard guidelines. We can get some concepts from the standard guidelines such as repair technical conditions, and we can analysis them.

(5) Fifth resource: periodical literature. There are often some repair knowledge in magazines such as engineering machine and repair, in part due to the strong flexibility of engineering equipment and its repair. So we can get some concept for conference.

(6) Sixth resource: related management information system. We can get some repair case from maintenance query system and maintenance management information system, and then we can construct case model.

3.3   Abstract Core Concepts and Built Hierarchy

On the basis of analysis and full collection of domain information, we list all the potential core concepts, and finally we confirmed eight core concepts by the way of identify, analysis and statistics, which include case, product, function, damage, phenomenon, environment, resource and disposal.

Case includes repair case and upkeep case, which mainly record history engineering equipment maintenance knowledge. Product is the aim of damage and maintain, and it is the basic object of damage mechanism and repair support countermeasure analysis. Function is an abstract description of the specifically ability of product or technology system, which depict the transport and conversion procedure of power stream, matter stream, and information stream, depict the efficacy and ability [4], we divide function into basic function and assistant function. Damage is the main cause of repair, we divide damage into battle damage, occasional failure, wear failure, unavailable supply, mis-operation, maladjusted and so on. Phenomenon is an important factor of fault diagnosis, we divide phenomena into physical phenomenon and biochemistry phenomenon. Physical phenomena include vision, smell, touch and hearing. We also can divide phenomenon into abnormal phenomenon and normal phenomenon. Environment is an important condition factor of maintenance decision-making. We divided it into geography and threaten, and further we divided element into highland, sea bells, Jack Frost, swamp, desert, woodland and plain, we divided hostility threaten into foreland battlefield and rear battlefield [5]. Resource is also an important factor of maintenance decision-making in battlefield or emergency. We divide resource into technique resource and entity resource, and further divide technique resource into repair tool, repair facility, repair equipment and repair personnel, and further divide technique resource into maintenance guide, upkeep regulation, maintain condition and so on. Disposal is a settle scheme of damage, which include using disposal and maintain disposal. Using disposal main include debase use, injured use, change operation mode and hazardous use. Maintain disposal main include upkeep and repair [6].

We expand core concepts, construct the concept model of whole ontology, and then we get the hierarchical model of concept, which is shown as Fig. 1.

3.4 Concept's Data Property Analysis

The hierarchical model of concept formed the main skeleton body of engineering equipment maintain ontology, but we should expand the concept according to the demand of description to complete the maintain domain ontology construction.

For example, we should use length, width, height, weight and material to describe product, and we will use type, characteristic, mechanism, cause, degree, disposal to describe the concept of damage, and we will use phenomena description and phenomena characteristic to describe phenomena, and we will use type, time, site, associated product, associated damage, description and evaluate to describe an case. We can get main data properties of engineering equipment maintain ontology by the way of analysis of the description demand of core concepts.

3.5 Concept's Object Property Analysis

According to the application of engineering equipment maintenance knowledge, we mainly analysis the knowledge from the perspective of the products' mechanism, join and dismounting, damage mode and maintain, which is specified as follows.

(1) Structure and Mechanism Relationship Analysis

It is very important to master the knowledge of engineering equipment structure and mechanism, which is basis to carry out maintenance support of engineering equipment. Reference to the FBS model, we think there are main seven object relationships, which include function hierarchical, function correlation, behavior contain, behavior cause and effect, structure hierarchical, structure generic, structure and function mapping [4].

(2) Join and Discounting Analysis

Replacement repair have been changed into the main repair means of basic-level troops in war. Join and discounting relationship of parts will influence the content and steps of replace, so we should analysis join and discounting relationship of parts on the basis of structure and mechanism. We divide assembling relationship into hierarchical, assort, connect, movement and constraint of assembly, and further we divide connect relationship into clearance fit, excessive cooperate, interference fit, and further we divide constraint into qualitative constraint and quantitative constraint. Fit, alignment, directional and insert relationships consist of qualitative constraint. Angle constraint and distance constraint consist of quantitative constraint.

(3) Analysis of Damage and Maintain Relationship

Damage location is very difficult in maintain of engineering equipment, because damage location associate with many knowledge, such as damage mode, damage mechanism, damage phenomena, damage characteristic, damage type, damage effect, damage dispose and damage case. The mainly relationship include cause and reason, correlation. We divide cause into direct cause and indirect cause, and we also divide cause into initial cause and final cause, we also divide reason into direct reason, indirect

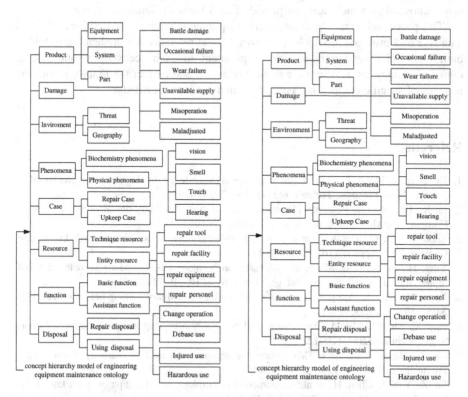

Fig. 1. Concept hierarchy model of engineering equipment maintenance ontology

Fig. 2. Object property model of engineering equipment maintenance ontology

reason, final reason. Correlation can also be divided into structure correlation and damage correlation [7].

On the basis of above analysis, we can summarize the relationships, then we get the main relationship include hierarchical relationship, assembling relationship and correlation, which is shown as Fig. 2.

3.6 Ontology Building, Application and Evolution

On the basis of concept hierarchy model, object property model, we tried to built the domain ontology of engineering equipment maintenance support with protege4.3. Ontology construction is an creative design processor. And domain ontology is developing with the study of domain. So the engineering equipment maintenance ontology will always evolve with all kinds of application ontology development [8].

4   Conclusion

Engineering equipment maintenance ontology is an system organization and expression for maintain knowledge of engineering equipment. We can share and reuse maintenance knowledge, which also can provide found basis for semantic search.

This paper preliminary built engineering equipment maintenance ontology with owl language and protege4.3, which can provide some reference for application ontology built on special version engineering equipment maintenance support and related domain ontology built. And we will further enrich and perfect the ontology with the analysis of specific engineering equipment maintenance knowledge.

References

1. Hu, J., Ji, L., Meng, Y., et al.: Equipment support knowledge ontology construction based on protege. Mod. Electron. Technol. 6(317), 207–210 (2010)
2. Li, K., Han, Z., Liu, P., et al.: The research and design of military domain ontology library. Comput. Knowl. Technol. 6(36), 10196–10198 (2010)
3. Zhou, Y., Li, Q.: Ontology modeling and semantic retrieval for aircraft fault knowledge. Comput. Eng. Appl. 47(16), 12–15 (2011)
4. Ying, H., Li, S., Guo, M., et al.: Research on ontology-based produce knowledge S-B-F representation model. Comput. Integr. Manuf. Syst. 10, 30–38 (2004)
5. Jiang, W., Hao, W., Yang, X.: Foundation of ontology in military training field. Comput. Eng. 34(5), 191–192 (2008)
6. Fan, X., Shi, C., Min, J.: Application and construction of ship gun maintenance. Armory Trans. Sichuan 32(8), 120–122 (2011)
7. Liu, L.: Research on the IETM of arms and equipments based on ontology. Northwestern Polytechnical University, pp. 53–59 (2007)
8. Su, Z., Yan, J., Chen, H., et al.: Construction of ontology-based equipment fault knowledge base. Syst. Eng. Electron. 37(9), 2067–2072 (2015)

Knowledge Graph Construction and Information Extraction

Boosting to Build a Large-Scale Cross-Lingual Ontology

Zhigang Wang(✉), Liangming Pan, Juanzi Li, Shuangjie Li, Mingyang Li,
and Jie Tang

Department of Computer Science and Technology,
Tsinghua University, Beijing 100084, People's Republic of China
{wzhigang,plm,ljz,lsj,lmy,tangjie}@keg.cs.tsinghua.edu.cn

Abstract. The global knowledge sharing makes large-scale multi-lingual knowledge bases an extremely valuable resource in the Big Data era. However, current mainstream Wikipedia-based multi-lingual ontologies still face the following problems: the scarcity of non-English knowledge, the noise in the multi-lingual ontology schema relations and the limited coverage of cross-lingual `owl:sameAs` relations. Building a cross-lingual ontology based on other large-scale heterogenous online wikis is a promising solution for those problems. In this paper, we propose a cross-lingually boosting approach to iteratively reinforce the performance of ontology building and instance matching. Experiments output an ontology containing over 3,520,000 English instances, 800,000 Chinese instances, and over 150,000 cross-lingual instance alignments. The F1-measure improvement of Chinese `instanceOf` prediction achieve the highest 32%.

Keywords: Ontology building · Instance matching · Cross-lingual

1 Introduction

As the Web is evolving to a highly globalized information space, sharing knowledge across different languages is attracting increasing attentions. Multilingual ontologies, in which the cross-lingual equivalent concepts or relationships are linked together using `owl:sameAs`, are important sources for harvesting cross-lingual knowledge from the Web and have significant applications such as multi-lingual information retrieval, machine translation and deep question answering. DBpedia [1], by extracting structured information from Wikipedia in 111 different languages, is a multi-lingual multi-domain knowledge base and becomes the nucleus of LOD. Obtained from WordNet and Wikipedia, YAGO, MENTA, and BabelNet are other famous large multi-lingual ontologies [6,7,12].

Though lots of researches have been done, there are still some problems to be solved. Firstly, the imbalance of different Wikipedia language versions leads to the highly unbalanced knowledge distribution in different languages. Figure 1 shows a simplified long tail distribution of the number of articles on

© Springer Nature Singapore Pte Ltd. 2016
H. Chen et al. (Eds.): CCKS 2016, CCIS 650, pp. 41–53, 2016.
DOI: 10.1007/978-981-10-3168-7_5

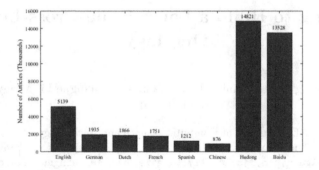

Fig. 1. Number of articles on major wikipedias, Hudong Baike and Baidu Baike

six major Wikipedia language versions. Most non-English knowledge in these ontologies is pretty scarce. Secondly, the noise of the large category system in Wikipedia leads to the incorrect semantic relations in these ontologies. For example, "Wikipedia-books-on-people is the `subCategoryOf` People" will lead to the wrong "Wikipedia-books-on-people is `subClassOf` People" in DBpedia's SKOS schema. And the relatively precise WordNet only cover some aspects of domains in English. Finally, because those ontologies are integrated directly by Wikipedia's cross-lingual links, the coverage of cross-lingual `owl:sameAs` relations in those ontologies is limited by the number of existing cross-lingual links.

On the other hand, there are more and more similar large-scale non-English online wikis in big data era. For example, the Chinese Hudong Baike and Baidu Baike, both containing more than 6 million articles, are even larger than the English Wikipedia (the largest Wikipedia language version). If multi-lingual ontology could be established between two large online wikis, such as English Wikipedia and Chinese Hudong Baike, multi-lingual ontologies with much higher coverage can be constructed.

In this paper, we try to build a large-scale cross-lingual ontology based on two heterogeneous online wikis in different languages. To our best of knowledge, we are the first to combine the processes of mono-lingual ontology building and cross-lingual instance matching together to build a cross-lingual ontology. Our work is motivated by two observations on the multilingual knowledge distributions. *Cross-lingual Knowledge Consistency*. A lot of facts are considered as correct all over the world, e.g. the facts about Science. Mining consistency across different languages not only helps to match equivalent cross-lingual knowledge, but also assists to improve the performance of mono-lingual ontology building each other. *Cross-lingual Knowledge Discordance*. The facts people concern or believe are quite different. E.g. the Chinese instance "China" is more linked to the Chinese locations but the English instance "China" is more linked to the counties in the world. Consideration of this problem in depth can help avoid incorrect matching.

This non-trivial task poses the challenges as follows, how to build two large-scale mono-lingual ontologies with correct semantic relations? How to construct an effective and efficient language-independent instance matching model? And how to boost the building of the cross-lingual ontology iteratively? Driven by these challenges, we propose a unified boosting framework to iteratively build a cross-lingual ontology. Our contributions are as follows.

1. We propose a binary classification-based method for large-scale mono-lingual ontology building, and a language-independent instance matching method. The ontology building method is able to eliminate the noise inside the wikis by predicting the correct `subClassOf` and `instanceOf` relations. The ontology matching method works for two highly heterogenous cross-lingual ontologies effectively and efficiently.
2. We propose a cross-lingually boosting method to reinforce the processes of ontology building and instance matching. The cross-lingual knowledge consistency and discordance are analyzed in depth. We iteratively expand the volume of labeled data for ontology building and expand the cross-lingual alignments for instance matching to improve the quality of built ontology simultaneously.
3. We conduct an experiment using the English Wikipedia and Hudong Baike data sets. Experimental results show that our boosting method outperforms the non-iterative method. The F1-measure of ontology building functions has an improvement of above 6%. In particular, the performance of Chinese `instanceOf` function get a high 32% improvement for F1-measure. A large ontology containing 3,520,000 English instances and 800,000 Chinese instances is built. Over 150,000 cross-lingual instance alignments are constructed.

2 Preliminaries

Basic Concepts. Given two online wikis in different languages and an initial aligment set, our target is to build two mono-lingual ontologies and find the equivalent alignments between them.

Definition 1. An **online wiki** is a graph containing a set of entities and a set of links between two entities. It can be formally represented as $\mathcal{G} = (\mathcal{V}, \mathcal{E})$, where $v \in \mathcal{V}$ denotes an *entity* and has an related *document*. We have $\mathcal{E} = \mathcal{V} \times \mathcal{V}$, and $e_{ij} \in \mathcal{E}$ indicate whether there exists a `subCategoryOf` or `articleOf`[1] relation from v_i to v_j (1 for yes, 0 for no).

Definition 2. An **ontology** is defined as a 2-tuple of the set of entities and the set of semantic relations. It can be formally represented as $\mathcal{O} = (\mathcal{X}, \mathcal{Y})$, where $x \in \mathcal{X}$ denotes a *concept* in the schema-level or an *instance* in the instance-level. We have $\mathcal{Y} = \mathcal{X} \times \mathcal{X}$ and $y_{ij} \in \mathcal{Y}$ indicate whether there exists a legal semantic relation from y_i to y_j (1 for yes, 0 for no). We only consider two kinds of semantic relations, which are `subClassOf` between two concepts and `instanceOf` from one instance to one concept.

---

[1] We use category and article to denote the concept and instance in the online wiki respectively.

Definition 3. The **alignment set** is the set of equivalent instances between two ontologies. It can be formally represented as $\mathcal{A} = \{a_i\}$, where $a_i = (x, x')$ denotes the equivalent instances between two ontologies respectively.

Problem Formulation. Given two online wikis $\mathcal{G}_1 = (\mathcal{V}, \mathcal{E})$, $\mathcal{G}_2 = (\mathcal{V}', \mathcal{E}')$ and an initial alignment set $\mathcal{A} = \{a_i\}_{i=1}^m$, we aim at constructing two mono-lingual ontologies $\mathcal{O}_1 = (\mathcal{X}, \mathcal{Y})$, $\mathcal{O}_2 = (\mathcal{X}', \mathcal{Y}')$ and a cross-lingual alignment set $\mathcal{A}' = \{a_i\}_{i=1}^n$. We have $n > m$, and $\mathcal{G}_1$, $\mathcal{G}_2$ are in two different languages[2]. The entities of the constructed ontologies are from the entities of online wikis, where $\mathcal{X} \subseteq \mathcal{V}$ and $\mathcal{X}' \subseteq \mathcal{V}'$. Thus, our major issue is to predict three kinds of relations, which are `subClassOf` between two concepts in each ontology, `instanceOf` from one instance to one concept in each ontology, and `equalTo` between two instances from two ontologies.

We formalize this problem as multiple binary classification problems. More formally, we are to learn two kinds of classification functions with a confidence output as follows.

- **Instance Matching Function** $f : \mathcal{X} \times \mathcal{X}' \mapsto [0, 1]$ to predict the probability to be `equalTo` relation between two instances x and x' from $\mathcal{O}_1$ and $\mathcal{O}_2$ respectively.
- **Ontology Building Function** $g_1 : \mathcal{V} \times \mathcal{V} \mapsto [0, 1]$ to predict the probability to be `subClassOf` or `instanceOf` relation between two entities v_i and v_j in $\mathcal{G}_1$, or $g_2 : \mathcal{V}' \times \mathcal{V}' \mapsto [0, 1]$ in $\mathcal{G}_2$.

To improve the performance of the isolated functions, we boost to mutually reinforce the learning of the building and matching functions.

3   Approach

As shown in Fig. 2, our approach is a boosting method. In each iteration we use the results of ontology building g_1, g_2 and instance matching f to reinforce the learning performance in the next iteration.

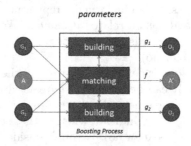

Fig. 2. Overview of the proposed approach

---

[2] We use $\mathcal{G}_1$ to represent the English online wiki, and use $\mathcal{G}_2$ to represent the Chinese online wiki.

3.1 Mono-lingual Ontology Building

We take the entities of $\mathcal{V}$, $\mathcal{V}'$ in the online wikis $\mathcal{G}_1$ and $\mathcal{G}_2$ as the entities of $\mathcal{X}$, $\mathcal{X}'$ in the ontoloties $\mathcal{O}_1$ and $\mathcal{O}_2$. Concretely, we take the categories in wikis as the concepts, and take the articles as the instances. Hence, our task is to learn the ontology building functions g_1 and g_2 to predict the correct subClassOf or instanceOf relations between two entities. We view both the correct subClassOf relation between two concepts and the correct instanceOf relation from an instance to a concept as an is-a relation. Table 1 shows some examples about the semantic relations generated from the online wikis.

Table 1. Examples of semantic relations

Entity 1	Relation	Entity 2	Right or wrong
European microstates	instanceOf	Microstates	Right
European microstates	instanceOf	Europe	Wrong
教育人物 (Educational Person)	subClassOf	人物 (Person)	Right
教育人物 (Educational Person)	subClassOf	教育 (Education)	Wrong

In this paper, we are to learn two series of functions $g_1 : \mathcal{V} \times \mathcal{V} \mapsto [0,1]$ and $g_2 : \mathcal{V}' \times \mathcal{V}' \mapsto [0,1]$ to predict the probabilities to be an is-arelation between two entities (1 for completely positive, 0 for completely negative). Notice that, we actually train four functions which are English subClassOf, English instanceOf, Chinese subClassOf an Chinese instanceOf, but we uniformly represent the ontology building functions of subClassOf and instanceOf in one language the same. The unique difference between them is that the input entities of subClassOf are two concepts but the input entities of instanceOf are one instance and one concept.

By manually labeling some training examples, we can learn the Logistic Regression models to get the ontology building functions g_1 and g_2. Table 2 shows the feature definition of g_1 function. The 10th feature is calculated as follows. We firstly list all of the sub-categories of current super-category. Then we calculate the frequency of each word in all of the sub-categories. The score of current sub-category is the sum of the frequency of each word in current sub-category. This feature is similar to a voting process, in which the more frequent words denote a higher probability. Similar as the 11th feature is.

The features in Table 2 are for learning the subClassOf predictor of g_1. The instanceOf features are similar, in which we replace the super-category into category and replace the sub-category into article. The head words can be extracted using a NLP parser. Note that, for features of g_2, we revise the 1st and 2nd features into "Is the sub-category starting with super-category" and "Is the sub-category ending with super-category" respectively. Besides, the basic unit for g_2 is one Chinese character but not a word. E.g. the 3rd feature is "the length of super-category characters".

Table 2. Feature definition for g_1

ID	Feature	Range
1	Is the head word of super-category plural?	$\{0, 1\}$
2	Is the head word of sub-category plural?	$\{0, 1\}$
3	Word length of super-category	Integer
4	Word length of sub-category	Integer
5	Word length of head words of super-category	Integer
6	Word length of head words of sub-category	Integer
7	Relation between the head words of super-category and sub-category	$\{\equiv, \subseteq, \supseteq, \perp, \triangle\}$
8	Does the non-head words of sub-category contain the head words of super-category?	$\{0, 1\}$
9	Does the non-head words of super-category contain the head words of sub-category?	$\{0, 1\}$
10	Score of sub-category	Numeric
11	Score of super-category	Numeric

$\equiv$ equivalent, $\subseteq$ smaller, $\supseteq$ larger, $\perp$ disjoint, $\triangle$ otherwise.

3.2 Cross-Lingual Instance Matching

Given the initial alignment set $\mathcal{A} = \{a_i\}_{i=1}^m$, cross-lingual instance matching is to generate a much larger alignment set $\mathcal{A}' = \{a_i\}_{i=1}^n$ ($n >> m$) between $\mathcal{O}_1$ and $\mathcal{O}_2$. We are to learn the function $f : \mathcal{X} \times \mathcal{X}' \mapsto [0, 1]$ to predict the probability to be equalTo relation between two instances x and x'.

By automatically sampling a part of alignments from $\mathcal{A}$ as the training examples, we can learn the Logistic Regression model to get the function f. We firstly present the features for instance matching, and then introduce two preprocessing methods, namely maximum clique pruning and link annotation. Finally, we present the post-processing method.

Feature Definition. The features used in f are designed by the observation of cross-lingual knowledge consistency. Both the lexical similarities and link-based structural similarities are defined. We use the following *Set Similarity* as the basic metric for structural similarities, which has been proven to be quite effective in [15]. Given two instances a and b, let S_a and S_b be their related sets of entities, the ***Set Similarity*** between a and b is calculated as

$$s(a, b) = \frac{2 \cdot |\phi_{1\to 2}(S_a \cap S_b)|}{|\phi_{1\to 2}(S(a))| + |S(b)|} \tag{1}$$

where $\phi_{1\to 2}(\cdot)$ maps the set of entities in $\mathcal{G}_1$ (or $\mathcal{O}_1$) to their equivalent entities in $\mathcal{G}_2$ (or $\mathcal{O}_2$) if the alignment exists.

Table 3 shows the feature definition of f. As we can see, both the structural similarities in the online wikis and in the ontologies are used.

Table 3. Feature definition for f

Type	ID	Feature	Description
Lexical	1	Edit-distance of titles without translation	Return 0 if there are no common characters
	2	Difference in word length	$\|English\_Word\_Length - Chinese\_Character\_Length\|$.
Structural	3	*Set Similarity* of categories	Calculated between $\mathcal{G}_1$ and $\mathcal{G}_2$
	4	*Set Similarity* of outlinks	Calculated between $\mathcal{G}_1$ and $\mathcal{G}_2$
	5	*Set Similarity* of inlinks	Calculated between $\mathcal{G}_1$ and $\mathcal{G}_2$
	6	*Set Similarity* of concepts	Calculated between $\mathcal{O}_1$ and $\mathcal{O}_2$

To overcome the link sparseness, we use a smoothing method in our experiments when computing those structural features.

Maximum Clique Pruning. Due to the cross-lingual knowledge discordance, the knowledge distributions across different languages differs a lot. Our feature definition is apt to choose the correspondences sharing more common related entities. However, we observe that a lot of neighbor entities are not very related in online wikis. E.g. in Hudong Baike, the article " "1月1日" " (1st, Jan.) is linked to many dates without much relatedness. This will lead to some erroneous correspondences such as " "1月1日" "(1st, Jan.) equalTo "3rd, May". We propose a maximum clique pruning to remove those structurally high linked but semantically low related structures. For each article in $\mathcal{G}_1$ or $\mathcal{G}_2$, we construct a local graph using this article and its linked articles. Then we calculate the maximum clique of this local graph. If the size of the maximum clique is larger 5, we prune the links between any two articles in the clique. In this way, lots of noise can be pruned from the online wikis. We add the similarities on the pruned network as new features for instance matching.

Link Annotation. Due to the link sparseness, the structural similarities across two heterogenous online wikis are quite sparse. To overcome this problem, we conduct a n-gram link annotation process to mine more links. The precision of link annotation is not sensitive, because we use the annotated links as new features for instance matching.

Heuristic Post-processing. Based on our observations, we propose the following rules to filter out some unreliable matching results: (1) *Multiple Correspondence*. If one English instance has been aligned to more than one Chinese instance, we remove all of those correspondences. (2) *Digits or Letters Co-occurrence*. If the Chinese instance's title contains a substring of more than two continuous digits or upper-case letters, we remove the correspondence if the English instance's title doesn't contain the same substring.

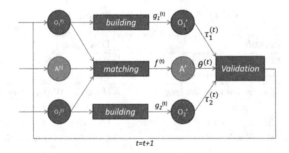

Fig. 3. Overview of boosting process in the iteration of t

3.3   Boosting to Build a Large-Scale Ontology

To boost a large-scale cross-lingual ontology, we iteratively learn the ontology building functions and the instance matching function. Figure 3 shows the overview of our boosting method in the iteration of t. Our boosting strategies are different for the building and matching functions.

Boosting the Ontology Building Process. The performance of ontology building functions is related to the volume of manually labeled data sets. Our idea is to expand the training data sets automatically after each iteration by using a cross-lingual semantic validation method. The detailed strategies are as follows.

- Train the ontology building functions $g_1^{(t)}$, $g_2^{(t)}$ using current training data sets.
- Predict the unlabeled data sets using the learned $g_1^{(t)}$, $g_2^{(t)}$.
- Validate the predicted data using current cross-lingual alignments as follows: if $f^{(t)}(x_1, x_1') > \theta^{(t)}$ and $f^{(t)}(x_2, x_2') > \theta^{(t)}$, then we have $g_1^{(t)}(x_1, x_2) = g_2^{(t)}(x_1', x_2') = 1$ if $g_1^{(t)}(x_1, x_2) + g_2^{(t)}(x_1', x_2') > (\tau_1^{(t)} + \tau_2^{(t)})$, and $g_1^{(t)}(x_1, x_2) = g_2^{(t)}(x_1', x_2') = 0$ if $g_1^{(t)}(x_1, x_2) + g_2^{(t)}(x_1', x_2') < (\tau_1^{(t)} + \tau_2^{(t)})$ (we experimentally set $\theta^{(t)}$, $\tau_1^{(t)}$ and $\tau_2^{(t)}$ to be 0.9, 0.5 and 0.5 respectively. A higher parameter value generates a stricter validation result).
- Expand the training data sets using the cross-lingually validated data.
- Iteratively repeat this process for the next iteration.

Boosting the Instance Matching Process. The structural features of instance matching process are calculated based on the initial alignment set. More alignments help to harvest more precise features. Thus, our idea is to expand the alignment set automatically after each iteration. The detailed strategies are as follows.

- Train the instance matching function $f^{(t)}$ using current alignments.
- Predict the unlabeled data sets using $f^{(t)}$.
- Validate the predicted data sets as follows: if $f^{(t)}(x, x') > \theta^{(t)}$, then we have $f^{(t)}(x, x') = 1$ (we experimentally set $\theta^{(t)}$ to be 0.9).
- Expand the alignment set using the validated alignments.
- Iteratively repeat this process for the next iteration.

4 Experiments

We conduct the experiments using English Wikipedia and Hudong Baike. The English Wikipedia dump is archived in August 2012, and the Hudong Baike dump is crawled from Huong Baike's website in May 2012. We remove all those entities in English Wikipedia, whose titles contain the following strings: *wikipedia, wikiprojects, lists, mediawiki, template, user, portal, categories, articles, pages, by*. We also remove the articles in Hudong Baike, which do not belong to any categories of Hudong. Table 4 shows the statistics of the cleaned online wikis.

Table 4. Statistics of cleaned data sets

Online Wiki	#Categories	#Articles	#Links	#Links/#Articles
English Wikipedia	561,819	3,711,928	63,504,926	17.1
Hudong Baike	28,933	980,411	23,294,390	23.8

Using the cross-lingual links between English and Chinese Wikipedias, we get an initial alignment set containing 126,221 alignments between English Wikipedia and Hudong Baike. We use Stanford Parser [2] for extracting the head words and use the Weka [3] toolkit for implementing the learning algorithms. We first evaluate the effectiveness of proposed mono-lingual ontology building and cross-lingual instance matching methods respectively, and then evaluate the proposed boosting approach as a whole.

4.1 Mono-Lingual Ontology Building

For the evaluation of mono-lingual ontology building, we randomly selected 3,000 English `subClassOf`, 1,500 Chinese `subClassOf`, 3,000 English `instanceOf`, and 1,500 Chinese `instanceOf` examples. We ask 5 graduate students of Tsinghua University to help us manually label those examples. The examples consented by more than 3 students are kept. Table 5 shows the detail of our labeled examples.

Table 5. Labeled data for mono-lingual ontology building.

Examples	subClassOf en	subClassOf zh	instanceOf en	instanceOf zh
Positive	2,123	780	2,097	638
Negative	787	263	381	518

en: English, **zh**: Chinese.

We conduct our experiments with a 5-fold cross-validation, and compare our Logistic Regression (LR) model with two baselines, namely Naïve Bayes (NB) and Support Vector Machines (SVM), using the same features defined in Sect. 3.1. As shown in Table 6, LR outperforms NB a lot and achieves comparative performance as the SVM (in most cases also outperforms SVM on F1-measure). In consideration of computation cost of the boosting process, our LR method is a good choice owing to its excellent learning efficiency.

Table 6. Results of mono-lingual ontology building. (%)

Methods	subClassOf en			subClassOf zh			instanceOf en			instanceOf zh		
	P	R	F1	P	R	F1	P	R	F1	P	R	F1
NB	87.1	62.5	72.8	87.1	85.9	86.5	95.8	42.7	59.1	60.2	55.7	57.9
SVM	80.8	86.7	83.6	83.8	98.6	**90.6**	84.5	100	91.6	53.1	82.1	64.5
LR	80.6	87.1	**83.7**	84.0	97.7	90.3	87.4	98.4	**92.6**	56.5	80.1	**66.3**

P: precision, R: recall, F1: F1-measure, **en**: English, **zh**: Chinese.

Table 6 also shows the cross-lingual performance comparison of `subClassOf` and `instanceOf` respectively. We find that English `instanceOf` performs better than Chinese `instanceOf`, but Chinese `subClassOf` is better than English `subClassOf`. This is because the 2nd and 3rd features in learning the building functions are linguistic related. The features are quite effective in learning English `instanceOf` and Chinese `subClassOf` respectively. That indicates the possibility to mutually improve the performance by the boosting process.

4.2 Cross-Lingual Instance Matching

In order to evaluate the cross-lingual instance matching method, we randomly select 3,000 initial alignments as the ground truth. We also automatically sample 10,000 random positive and 25,000 random negative alignments as the training data sets. In the experiments, we aim to investigate how the instance matching method performs before and after the heuristic post-processing (**HP**), and how the instance matching performs with different numbers of alignments. Therefore, we conduct four groups of experiments, each of which uses different number of alignments. In each group, we also compare the performance of our method before and after the heuristic post-processing. Table 7 shows the detailed results. The precision of our method is relatively high but the recall is rather low. We think this still works for our boosting method because the recalled alignments can be enriched iteratively even the recall is relatively low. However, a low precise alignment results will deteriorate the boosting process rapidly.

Table 7. Results of cross-lingual instance matching. (%)

#Alignments	Before HP			After HP		
	Precision	Recall	F1-measure	Precision	Recall	F1-measure
0.03 Mil	81.5	5.6	10.5	91.4	5.6	10.6
0.06 Mil	86.4	6.0	11.3	91.9	6.0	11.3
0.09 Mil	**89.7**	6.5	12.0	**93.9**	6.5	12.2
0.12 Mil	86.5	6.8	**12.6**	88.9	6.8	**12.6**

As we can see from Table 7, in each group of the experiments, our method always performs better after the heuristic post-processing (especially for the precision). It shows the heuristic post-processing method can effectively filter out the unreliable matching results. On the other side, the F1-measure of our approach always increases when more alignments are used. Therefore, expanding the initial alignment set iteratively is important for improving the instance matching performance.

4.3 Boosting to Building a Large-Scale Ontology

At last, we evaluate our approach as a whole. For ontology building, we use the same labeled data sets and iteratively boost our approach. Table 8 shows that the performance of the four ontology building functions increases in each iteration. In particular, the precision and recall of Chinese `instanceOf` function goes from 65.0% and 63.0% to 96.7% and 96.9% respectively. As we can see, the performance after three iterations is excellent.

Table 8. Results of boosting to build the ontology. (%)

Iteration	subClassOf en			subClassOf zh			instanceOf en			instanceOf zh		
	P	R	F1	P	R	F1	P	R	F1	P	R	F1
Iteration 1	80.8	88.2	84.4	82.0	100	90.1	87.4	97.1	92.0	65.0	63.0	64.0
Iteration 2	87.3	91.8	89.5	91.8	98.6	95.1	93.3	98.4	95.8	91.4	89.1	90.2
Iteration 3	87.7	93.4	**90.5**	94.8	99.3	**97.0**	97.3	99.6	**98.4**	96.7	97.0	**96.9**

P: precision, R: recall, F1: F1-measure, **en:** English, **zh:** Chinese.

In our experiments, we stop after the third iteration and successfully get two ontologies as shown in Table 9. For ontology matching, we use the same training data sets and all of the 126,221 alignments as the initial alignment set. We iteratively repeat the boosting process and 31,108 new alignments are found after 100 iterations. Due to the high computation cost, more iterations are still ongoing to find more alignments.

Table 9. Results of built ontology

	#Concepts	#Instances	#subClassOf	#instanceOf
English	479,040	3,520,765	751,154	11,339,698
Chinese	24,243	803,278	29,655	2,144,000

5 Related Work

Multi-lingual Ontology Building. Ontology building is to generate an ontology concerning some specific domains in the form of Resource Description Framework. Current ontology building strategies can be grouped into three categories, namely manual construction, crowdsourcing based approach [13] and open Web extraction approach. The costly manual constructed ontologies, such as Word-Net, HowNet and Cyc, are relatively high-quality but usually only cover parts of facts and are costly to maintain. Crowdsourcing based approach is becoming a prevalent method for building a large-scale and regularly updated ontology. DBpedia, by making the Wikipedia machine-readable, is a representative of this approach [1]. YAGO [12], MENTA [6] and BabelNet [7] are other multi-lingual ontologies based on WordNet and Wikipedia. Zhishi.me [8] is a Chinese knowledge base by integrating Hudong Baike, Baidu Baike and Chinese Wikipedia. XLORE [16] is a multilingual ontology generated from Hudong Baike, Baidu Baike, Chinese Wikipedia and English Wikipedia. Ponzetto and Strube have proposed some methods based on connectivity in the network and lexico-syntactic matching to derive a taxonomy from Wikipedia [9]. The open Web extraction approach aims to find a wider range of knowledge in the Web. This method gives us more opportunities to harvest more knowledge, but involves more noise and need to build an ontology from scratch. Probase [17] and TextRunner [18] are representatives of open Web extraction approach. Our proposed approach is a crowdsourcing based cross-lingual ontology building method.

Cross-lingual Ontology Matching. Ontology matching is to find equivalent correspondences between semantically related entities of ontologies [4,11]. Current ontology matching strategies can be grouped into two categoies, namely heuristic-based approach and machine learning-based approach. By manually defining some weights or threshold values, such heuristic-based approaches as similarity flooding and similarity aggregation can resolve the ontology matching problem quite efficiently and effectively. RiMOM [5] is a multi-strategy ontology alignment framework. The machine learning-based approach is to learn the weights and threshold values automatically. Rong et al. have proposed a transfer learning-based binary classification approach for instance matching [10]. Wang et al. have proposed a linkage factor graph model to match the instances across heterogenous wiki knowledge bases [15]. Current cross-lingual ontology matching approaches usually employ a generic two-step method, where ontology labels are translated into the target natural language first and monolingual matching techniques are applied next [5,14]. Wang et al. proposed a language-independent linkage factor graph model for instance matching [15]. Our proposed approach is a classification-based language-independent boosting method.

6   Conclusion and Future Work

In this paper, we propose a boosting method to build a large-scale cross-lingual ontology. The performance of ontology building and instance matching is reinforced iteratively. In particular, the performance of Chinese `instanceOf` function get a high 32% improvement for F1-measure. In our future work, we will iteratively find more cross-lingual instance alignments and crawl more Hudong Baike articles to enrich the Chinese instances. We will also improve our cross-lingual instance matching model to improve the recall, which is relatively low currently.

References

1. Auer, S., Bizer, C., Kobilarov, G., Lehmann, J., Cyganiak, R., Ives, Z.: DBpedia: a nucleus for a web of open data. In: Aberer, K., Choi, K.-S., Noy, N., Allemang, D., Lee, K.-I., Nixon, L., Golbeck, J., Mika, P., Maynard, D., Mizoguchi, R., Schreiber, G., Cudré-Mauroux, P. (eds.) ASWC/ISWC -2007. LNCS, vol. 4825, pp. 722–735. Springer, Heidelberg (2007). doi:10.1007/978-3-540-76298-0_52
2. Green, S., de Marneffe, M.C., Bauer, J., Manning, C.D.: Multiword expression identification with tree substitution grammars: a parsing tour de force with french. In: EMNLP (2011)
3. Hall, M., Frank, E., Holmes, G., Pfahringer, B., Reutemann, P., Witten, I.H.: The WEKA data mining software: an update. SIGKDD 11, 10–18 (2009)
4. Jean-Mary, Y.R., Shironoshita, E.P., Kabuka, M.R.: Ontology matching with semantic verification. Web Semant. 7, 235–251 (2009)
5. Li, J., Tang, J., Li, Y., Luo, Q.: RiMOM: a dynamic multistrategy ontology alignment framework. TKDE 21, 1218–1232 (2009)
6. de Melo, G., Weikum, G.: MENTA: inducing multilingual taxonomies from wikipedia. In: CIKM (2010)
7. Navigli, R., Ponzetto, S.P.: BabelNet: the automatic construction, evaluation and application of a wide-coverage multilingual semantic network. Artif. Intell. 193, 217–250 (2012)
8. Niu, X., Sun, X., Wang, H., Rong, S., Qi, G., Yu, Y.: Zhishi.me - weaving Chinese linking open data. In: Aroyo, L., Welty, C., Alani, H., Taylor, J., Bernstein, A., Kagal, L., Noy, N., Blomqvist, E. (eds.) ISWC 2011. LNCS, vol. 7032, pp. 205–220. Springer, Heidelberg (2011). doi:10.1007/978-3-642-25093-4_14
9. Ponzetto, S.P., Strube, M.: Deriving a large scale taxonomy from wikipedia. In: AAAI (2007)
10. Rong, S., Niu, X., Xiang, E.W., Wang, H., Yang, Q., Yu, Y.: A machine learning approach for instance matching based on similarity metrics. In: Cudré-Mauroux, P., Heflin, J., Sirin, E., Tudorache, T., Euzenat, J., Hauswirth, M., Parreira, J.X., Hendler, J., Schreiber, G., Bernstein, A., Blomqvist, E. (eds.) ISWC 2012. LNCS, vol. 7649, pp. 460–475. Springer, Heidelberg (2012). doi:10.1007/978-3-642-35176-1_29
11. Shvaiko, P., Euzenat, J.: Ontology matching: state of the art and future challenges. TKDE 25, 158–176 (2013)
12. Suchanek, F.M., Kasneci, G., Weikum, G.: Yago: a core of semantic knowledge. In: WWW (2007)
13. Tang, J., Leung, H.f., Luo, Q., Chen, D., Gong, J.: Towards ontology learning from folksonomies. In: IJCAI (2009)
14. Trojahn, C., Quaresma, P., Vieira, R.: A framework for multilingual ontology mapping. In: LREC (2008)
15. Wang, Z., Li, J., Wang, Z., Tang, J.: Cross-lingual knowledge linking across wiki knowledge bases. In: WWW (2012)
16. Wang, Z., Li, J., Wang, Z., Li, S., Li, M., Zhang, D., Shi, Y., Liu, Y., Zhang, P., Tang, J.: XLore: A large-scale English-Chinese bilingual knowledge graph. In: ISWC (2013)
17. Wu, W., Li, H., Wang, H., Zhu, K.Q.: Probase: a probabilistic taxonomy for text understanding. In: SIGMOD (2012)
18. Yates, A., Cafarella, M., Banko, M., Etzioni, O., Broadhead, M., Soderland, S.: Textrunner: open information extraction on the web. In: NAACL-Demonstrations (2007)

Large Scale Semantic Relation Discovery: Toward Establishing the Missing Link Between Wikipedia and Semantic Network

Xianpei Han[✉], Xiliang Song, and Le Sun

State Key Laboratory of Computer Sciences, Institute of Software,
Chinese Academy of Sciences, Beijing 100190, China
{xianpei,xiliang,sunle}@nfs.iscas.ac.cn

Abstract. Wikipedia has been the largest knowledge repository on the Web. However, most of the semantic knowledge in Wikipedia is documented in natural language, which is mostly only human readable and incomprehensible for computer processing. To establish the missing link from Wikipedia to semantic network, this paper proposes a relation discovery method, which can: (1) discover and characterize a large collection of relations from Wikipedia by exploiting the *relation pattern regularity*, the *relation distribution regularity* and the *relation instance redundancy*; and (2) annotate the hyperlinks between Wikipedia articles with the discovered semantic relations. Finally we discover *14,299* relations, *105,661* relation patterns and *5,214,175* relation instances from Wikipedia, and this will be a valuable resource for many NLP and AI tasks.

Keywords: Semantic network · Relation discovery · Knowledge acquisition

1 Introduction

A long-standing goal of natural language processing (NLP) and artificial intelligence (AI) is to build large-scale, machine-readable knowledge base (KB) which can support natural language understanding and human-like reasoning. To achieve this goal, a continuum of research, from the manual construction to the automatic information extraction, have been devoted to the knowledge base construction. In its early stages, researchers attempted to build KBs by manually collecting common sense knowledge. Several most notable examples include, *WordNet* (Miller 1995), *FrameNet* (Baker et al. 1998) and *OpenCyc* (Matuszek et al. 2006). These manually constructed methods, however, require too much manual engineering and are not suitable for constructing high-coverage and up-to-date KBs which fit to the real world usage.

   To overcome the limitations of manually constructed KBs, there have been many research efforts devoted to the fully automatic open information extraction (Open IE) techniques, which extract facts (i.e., relational tuples such as *Headquarters-In(Armonk, IBM)*) from a large corpus or web in a *Bootstrapping* or *self-learning* way. Several notable examples include *DIPRE* (Brin 1999), *Snowball* (Agichtein and Gravano 2000), *KnowItAll* (Etzioni et al. 2004), *TextRunner* (Yates et al. 2007) and *NELL* (Carlson et al. 2010). These open IE methods, however, often fail in achieving the high

© Springer Nature Singapore Pte Ltd. 2016
H. Chen et al. (Eds.): CCKS 2016, CCIS 650, pp. 54–66, 2016.
DOI: 10.1007/978-981-10-3168-7_6

quality due to the limited performance of automatic IE techniques. For instance, only 1 million (9.1%) of the 11 million relation tuples extracted by *TextRunner* were concrete facts (Banko, Cararella et al. 2007).

In recent years, Wikipedia provides a *large-scale*, *structure-rich* and *up-to-date* text corpus, which contains more than 4,000,000 articles and with rich semantic structures such as *Categories*, *Links* and *Infoboxes*. Wikipedia provides a new opportunity for knowledge base construction. Unfortunately, the task of harvesting semantic knowledge from Wikipedia (and other knowledge sharing sites) is challenging. That is, in spite of its rich structure (e.g., each *company* has its own article, and has links to its *products, headquarter, founder, CEO,* et al.), Wikipedia contents are still mostly only human readable. Semantic knowledge, e.g., the semantic relation between concepts, is not *formally* and *explicitly* stated. For example, although the article *IBM* contains links to *Thomas J. Watson, Thomas Watson Jr.* and *1911*, the semantic of these links are implicitly stated in natural language sentences such as "*The company was founded in 1911 by Thomas J. Watson*". Based on the above observation, we believe that there is *a missing link* between Wikipedia and a machine processable semantic network, i.e., the meaning of links is documented in natural language only – a representation which is incomprehensible for computer processing and its meaning is unclear to computer, therefore most knowledge in Wikipedia cannot be directly used in common sense reasoning and natural language understanding.

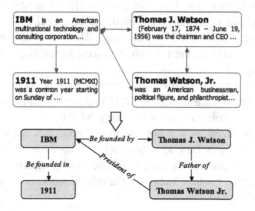

Fig. 1. Currently there are articles and links (above), our system will finally extract semantic relations from the links between these arguments (below).

In this paper, we want to establish the missing link from Wikipedia to a semantic network by providing a formalized, machine-processable semantic definition for the links between Wikipedia articles, see Fig. 1 as an example. To achieve this goal, this paper proposes a semantic relation discovery method, which can:

(1) Discover and characterize a large collection of semantic relations, which propose a formalized way to define the semantic of links; and

(2) Annotate the links between Wikipedia articles using the above set of semantic relations.

Specifically, our method extracts relation patterns and discover semantic relations by exploiting the regularity and the redundancy of semantic relations:

(1) **Regularity**: Although there is nearly unlimited ways to express a specific relation, in many cases basic principles of economy of expression and/or conventions of genre will ensure that certain systematic ways (i.e., the patterns) will be used to express a specific relation (Wang et al. 2012). For example, in most cases the *IS-A* relation will be expressed by the pattern *"Arg1 is a Arg2"*, although there may exist many other ways to express it. This paper refers this regularity as *relation pattern regularity*. Based on the relation pattern regularity, we believe that the patterns of relations will be repeatedly used to the extent that it can be identified and categorized from a large corpus.

(2) **Redundancy**. Due to the regularity and the large size of Wikipedia, the same relation instance will be expressed redundantly in many different ways and many times. For example, the relation *Be-Founded-In(IBM, 1911)* is expressed in many different ways in Wikipedia such as the *link between IBM and 1911*, the *Infobox of IBM*, and natural language sentences such as *"IBM was founded in 1911"*. This paper refers this redundancy as *relation instance redundancy*.

Based on the above observations, we propose to exploit the above regularity and redundancy using a hierarchical Dirichlet process (HDP) model (Teh et al. 2006), where the regularity and redundancy are modeled as statistical distributions and the dependencies between them. Furthermore, the HDP can adaptively determine the number of relations underlying the relation instances, which is a challenging problem for relation discovery.

We have applied our relation discovery method to Wikipedia, and finally *14,299* relations, *105,661* relation patterns and *5,214,175* relation instances are discovered. We believe this will be a valuable resource for many NLP tasks.

This paper is organized as follows. Section 1 describes the data preprocessing step. Section 2 demonstrates how to extract relation instances from Wikipedia. Section 3 describes how to discover semantic relations using the HDP model. Section 4 presents the experiments. Section 5 reviews the related work. Section 6 concludes this paper.

2 Data Preprocessing

In this section, we describe the data preprocessing steps for Wikipedia, including Wikipedia text preprocessing and entity linking.

2.1 Wikipedia Text Preprocessing

In this paper, we use the Jan. 30, 2010 English version of Wikipedia. Given the Wikipedia data, we first segment the main content of each article into sentences, and discard the sentences which are too short (<4 words) or too long (>50 words). Finally we collect 26,852,307 sentences. For each sentence, we tokenize, tag and parse them using the Stanford CoreNLP Tools[1].

2.2 Entity Linking

In order to discover semantic relations between entities, we need to identify all occurrences of a specific entity. Unfortunately, there are many different ways to mention a specific entity, including *name mentions*, *nominal mentions* and *pronoun mentions* (Doddington et al. 2004). For example, the company *IBM* may be mentioned by its name *IBM*, the nominal *the company* and the pronoun *it*.

To resolve the above problem, this paper links all mentions with their referent entities (i.e., *entity linking*) through the following two steps:

Linking Name Mentions to Entities. In this step, we link all name mentions to their referent entity. There have been a lot of entity linking methods, in this paper we use the entity-topic model described in (Han and Sun 2012), which collectively links all name mentions in a document by exploiting both the mention context and the document topics.

Linking Subject Mentions to Entities. In this step we link the nominal mentions and the pronoun mentions to their referents (e.g., it → IBM). In this paper, we use the method described in Li et al. (2010), which identify the subject mentions of a Wikipedia article by finding the top 3 frequent subject noun phrases of a Wikipedia article.

2.3 Relation Instance Extraction

This section describes how to extract the relation instances from Wikipedia. Given a pair of entities in a sentence, then we describe how to: (1) extract the phrase in the sentence which expresses the relation between them; and (2) validate whether the extracted phrase is a relation pattern based on the relation instance redundancy and the relation pattern regularity.

Relation Phrase Extraction. In this paper, a relation phrase is the phrase in a sentence which expresses the relation information between two given entities. For example, the relation phrase for entities *IBM* and *1911* in sentence "*IBM was founded in 1911 by Thomas J. Watson.*" should be "*IBM was founded in 1911*".

According to Bunescu and Mooney (2005), most of the information for identifying a relation between two entities is in the *shortest dependency path* (*SDP*) between them. Furthermore, we also observed that some modifiers of the SDP words also contain the relation information about the two entities. For example, in Fig. 2, the *auxpass* modifier *was* of the SDP word *founded* is also useful for expressing the relation between (*IBM*, *1911*) and between (*IBM*, *Thomas J. Watson*).

---

[1] http://nlp.stanford.edu/software/corenlp.shtml.

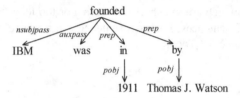

Fig. 2. A typed dependency parse tree

Based on the above observation, given two entities in a sentence, this paper extracts the relation phrase of them as follows:

First, we extract all words in the SDP between the two arguments as relation phrase. For instance, in Fig. 2 the SDP words *"IBM founded in 1911"* is identified for *(IBM, 1911)*;

For each SDP word, we add the selected modifiers of them to the relation phrase. Using the Stanford's typed dependencies (Marneffe and Manning 2008), the modifiers we used for SDP words are shown in Table 1. For instance, in Fig. 2 the modifier word will be added to the relation phrase for *(IBM, 1911)*, and now the relation phrase is *"IBM was founded in 1911"*.

Table 1. Selected modifiers for SDP word

POS	Typed modifiers
Verb	*aux, auxpass, cop, neg*
Noun	*cop, det, neg*

Relation Instance Extraction. Through the relation phrase extraction, we can extract many relation phrases. For example, in Fig. 2 we can identify three entity pairs corresponding with their relation phrases:

- (IBM, 1911): *Arg1 be found in Arg2*
- (IBM, Thomas J. Watson): *Arg1 be found by Arg2*
- (1911,Thomas J. Watson): *be found in Arg1 by Arg2*

Unfortunately, not all relation phrases are relation patterns, e.g., the phrase "be found in Arg1 by Arg2" in above. Therefore, we need to filter out noisy relation instances. In this paper, we filter out noisy relation instances using three constraints:

Syntactic Constraint. As shown in (Etzioni et al. 2008) and (Chan and Roth 2011), the relation pattern usually follows some specific syntactic patterns. Therefore we can filter out the relation phrases which are not consistent with these syntactic patterns. In this paper, we assume that all relation patterns should be consistent with the Verb patterns in (Chan and Roth 2011), i.e., the two arguments should head in the same verb, with one argument the subject of the head verb, and the other argument the object or the preposition object of the head verb.

Link Constraint. Based on the relation instance redundancy, a relation instance should occur in many different ways. Therefore, we can filter out the relation instances which occur in only one way. In Wikipedia, a link between two articles usually indicates the existence of semantic relation between them, therefore we can filter out the relation instances with no link between their arguments. For example, if there is no link between the articles *1911* and *Thomas J. Watson*, we will filter out all relation instances whose arguments are (*1911, Thomas J. Watson*).

Significance Constraint. Based on the relation pattern regularity, a relation pattern will be used frequently to express a specific relation. For example, the "*Arg1 be a Arg2*" will be used many times to express the *IS-A* relation. Based on this observation, we filter out all relation phrases whose occurrences are below a specific threshold (5 times in this paper).

Using the above three constraints, our method finally identifies *105,661* relation patterns and *5,214,175* relation instances. Table 2 demonstrates the top five frequent relation patterns extracted from Wikipedia.

Table 2. The top 5 frequent relation phrases

Relation pattern	Frequency
Arg1 be a Arg2	679,081
Arg1 be Arg2	234,081
Arg1 have Arg2	74,266
Arg1 became Arg2	39,628
Arg1 be born in Arg2	36,390

2.4 Argument Classification

Finally, we add the argument type information to the relation instance. Although Wikipedia has a category system, its categories are mostly thematic facets (Ponzetto and Navigli 2009) rather than categories from a well-formed taxonomy. For example, the article *IBM* is labeled with categories "*Companies listed on the New York Stock Exchange*", "*1911 establishments in the United States*", etc. To resolve the above problem, this paper uses WordNet as the taxonomy and labels each argument with a WordNet synset using the method described in (Ponzetto and Navigli 2009).

Through the above relation phrase extraction, relation instance extraction and argument classification steps, we extract and represent each relation instance as a 5-tuple (*Arg1, Arg1 Type, Arg2, Arg2 Type, Relation Pattern*). For example, the relation instance *Be-Founded-In(IBM, 1911)* will be represented as (*IBM, Company, 1911, Year, Arg1 be found in Arg2*).

3 Discovering Semantic Relations Using HDP Model

In this section, we describe how to discover and characterize a large collection of semantic relations from the extracted relation instances. Specifically, we address three problems in this section:

(1) How many different underlying semantic relations for the extracted relation instances?
(2) How to represent and characterize the discovered semantic relations?
(3) For each relation instance, which semantic relation it expresses?

As described in Sect. 1, we resolve the above problems based on the idea that: (1) *Relation Pattern Regularity*, i.e., a certain systematic patterns will be used to express a specific relation; and (2) *Relation Distribution Regularity*, i.e., the relations for each argument type pair are usually selected from a regular and fixed set and follow a specific distribution. Based on the above idea, then we propose to model and exploit them using a hierarchical Dirichlet process model (HDP).

3.1   Document and Relation Representation

Based on the relation distribution regularity, we organize all relation instances with the same argument types into an individual document, so that the patterns in the same document will have a high likelihood to be assigned to the same relation. For example, Fig. 3 shows a document for the argument type pair (*Actor, Actor*), corresponding with their relation patterns' count.

Doc: *Actor-Actor*	
Arg1 be a Arg2	*1,480*
Arg1 appear with Arg2	*583*
Arg1 star with Arg2	*519*
Arg1 be married to Arg2	*471*
Arg1 marry Arg2	*440*

Fig. 3. A demo of the *Actor-Actor* document

Based on the relation pattern regularity, we model each relation as a multinomial distribution of relation patterns. Figure 4 demonstrates the learned pattern distribution of the well-known *IS-A* relation.

Relation: *IS-A*	
Arg1 be a Arg2	*0.940*
Arg1 be establish as Arg2	*0.009*
Arg1 be among Arg2	*0.008*
Arg1 be consider one of Arg2	*0.005*
Arg1 be seen as Arg2	*0.005*

Fig. 4. The top 5 patterns of the *IS-A* relation

3.2 Hierarchical Dirichlet Process Model

In this section, we describe how to exploit the redundancy and the regularity using a Hierarchical Dirichlet Process (HDP) model. Specifically, the HDP model assumes that all documents are generated through the following process (Teh et al. 2006):

1. Draw the corpus level (global) relation distribution $\beta \sim$ GEM(γ). For example, in Fig. 5 the corpus probabilities for the three relations may be drawn as $\beta = \{Appear\text{-}With^{0.3}, IS\text{-}A^{0.4}, Be\text{-}Acquire\text{-}By^{0.3}\}$;
2. For each relation $z \in \{1, 2, \ldots\}$, draw its relation pattern distribution $\phi_z \sim$ Dirichlet(δ);
3. For each document d_j (i.e., a specific argument type pair), draw the document's specific relation distribution $\pi_i \sim DP(\alpha, \beta)$. For instance, in Fig. 5 we may draw the relation probabilities for document *(Actor, Actor)* as $\{Appear\text{-}With^{0.6}, IS\text{-}A^{0.4}\}$, and for document *(Company, Company)* as $\{IS\text{-}A^{0.3}, Be\text{-}Acquire\text{-}By^{0.7}\}$.
4. For each relation instance x_i in a document d_j:
   (a) Draw the expressed relation of the instance xi as $z_i \sim \pi_j$;
   (b) Draw the relation pattern from the pattern distribution of relation z_i as $x_i \sim \phi_{zi}$.

In HDP model, the pattern distributions of the same relation are shared across all documents, therefore the relation pattern regularity can be exploited, i.e., the same pattern distribution will be used in all documents. For example, in Fig. 5 the pattern distribution of the IS-A relation will be shared across the docs (Actor, Actor) and (Company, Company). Furthermore, for each document, their relation distribution is draw from the corpus relation distribution with a concentration parameter α. Thus the HDP will put a concentrated relation distribution for each document, and the relation distribution regularity can be modeled by selecting an appropriate α. For example, although there are three global relations, only two of them will appear in doc *(Actor, Actor)*.

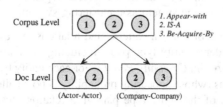

Fig. 5. A demo of the relation distribution generation of HDP

The Inference of HDP. As the same as (Teh et al. 2006), the Gibbs sampler for the HDP in this paper is as follows:

$$
\begin{cases}
z_{ji} = t & \alpha \quad \dfrac{n_{jt}^{-1}}{n_{j*}^{-i} + \alpha} f(x_{ji}, \phi_t) \\[2ex]
z_{ji} = t^g & \alpha \quad \dfrac{\alpha}{n_{j*}^{-i} + \alpha} \dfrac{m_{tg}^{-i}}{m_*^{-i} + \gamma} f(x_{ji}, \phi_{tg}) \\[2ex]
z_{ji} = t^{new} & \alpha \quad \dfrac{\alpha}{n_{j*}^{-i} + \alpha} \dfrac{\gamma}{m_*^{-i} + \gamma} f(x_{ji}, \phi_{t^{new}})
\end{cases}
$$

where x_{ji} is the ith instance in d_j, z_{ji} is relation assignment for x_{ji}, $z_{ji} = t$ means assign x_{ji} with a relation which has appeared in document d_j, and $z_{ji} = t^g$ means assign x_{ji} with a relation has appeared in corpus, and $z_{ji} = t^{new}$ means assign x_{ji} with a new relation, n_{jt}^{-i} and m_t^{-i} correspondingly the appearance count of relation t in document d_j and in corpus, $n_{j*}^{-i} = \sum_t n_{jt}^{-i}$, $m_*^{-i} = \sum_t m_t^{-i}$, the $f(x_{ji}, \phi_t)$ is the likelihood of generating pattern x_{ji} from relation t.

Notice that the above HDP model is a non-parametric Bayesian model, it can generate new relation when $z_{ji} = t^{new}$ is sampled, therefore it can adaptively determine the number of relations underlying the extracted relation instances. Furthermore, the above inference process can identify which semantic relation a relation instance expressed by assigning it with a relation. After the assignment, we can easily get the pattern distribution of all relations by estimating them from the final assignments.

The Hyperparameter Setting. In HDP model, the hyperparameter α controls the number of relations in a document, and α together with γ control the number of relations in a corpus. In this paper, following Escobar and West (1995), we put vague gamma priors on α and γ so that their values can be adaptively learned. Concretely, we set $\alpha \propto \text{Gamma}(1, 1)$ and $\gamma \propto \text{Gamma}(1,1)$, and the final value of α and γ in our data set are correspondingly around *1790* and *27*. For σ, we set σ to a small value *0.0001* so that HDP can express a relation with a regular and fixed set of patterns.

4 Experiments

4.1 Experimental Settings

Generally, the semantic relation discovery is a process of grouping relation patterns into clusters $C = \{C_1, C_2, ..., C_n\}$, with each cluster C_i representing a semantic relation. Therefore, we can evaluate the system as a clustering system.

Data Set. Due to the size of relation patterns and relations, we evaluate the quality of discovered relations under 4 entity type pairs, including *Company-Month*, *Company-Company*, *Company-City* and *Company-People*. For each entity type pair, we manually group the salient patterns (whose appearing probability is no smaller than 5% in at least one discovered relation of the entity type pair) into relation clusters $L = \{L_1, L_2, ..., L_m\}$, with each relation cluster is a set of relation patterns indicating the same relation.

Evaluation Criteria. Given the discovered relations C and the manually clustered relations L, we evaluate the quality of the discovered semantic relations using standard clustering metrics: *Purity, Inverse Purity* and *F-Measure* (Amigo et al. 2009).

Baselines. We compare our method with two baselines:

(1) ***One_in_One:*** The first is *One_in_One*, which assigns each relation pattern to an individual cluster, therefore the Purity of *One_in_One* will always be *1.0*.
(2) ***ET_in_One***: The second is *ET_in_One*, which assigns all relation patterns with same entity argument types into a single cluster. In our data set the Inverse Purity of *ET_in_One* will always be 1.0.

4.2   Experimental Results

In this section we demonstrate and discuss the experimental results. Table 3 shows the size of discovered relations and Table 4 shows the quality of the discovered relations.

Table 3. The size of discovered semantic relations

Relation	Relation pattern	Relation instance
14,299	105,661	5,214,175

Table 4. The quality of discovered relations

	Pur	Pur_Inv	F
One_in_One	1.00	0.29	0.45
ET_in_One	0.40	1.00	0.57
Our method	0.77	0.56	0.65

From the Tables 3 and 4, we can see that:

(1) Our method can discover a large collection of relations: totally *14,299* relations, *105,661* patterns and *5,214,175* instances are discovered. We believe this will be a valuable resource for many NLP tasks;

(2) Our method can discover homogeneous and complete relations: the average *Purity* and *Inverse Purity* of learned relations are about 0.77 and 0.56, and a 20% and 8% F-measure improvements are achieved over the *One_in_One* and the *ET_in_One* baselines. This means that for each resulting cluster around 77% patterns within it will express the same relation, and for each relation there will be a cluster which can capture around 56% patterns of it.

Table 5. Some examples of learned relations

Relation	Top 5 frequent patterns with prob.	
(Company, Month)#1	*Arg1 be found on Arg2*	0.391
	Arg1 be incorporate on Arg2	0.126
	Arg1 be found Arg2	0.076
	Arg1 be form in Arg2	0.067
	in Arg1, Arg2 merge to form	0.047
(Company, Company)#1	*Arg1 be sold to Arg2*	0.311
	Arg1 be acquire by Arg2	0.248
	Arg1 acquire in Arg2	0.078
	Arg1 own Arg2	0.039
	Arg1 be list a constituent of Arg2	0.031
(Company, City)#1	*Arg1 be headquarter in Arg2*	0.280
	Arg1 establish in Arg2	0.139
	Arg1 be establish in Arg2	0.099
	Arg1 be open in Arg2	0.037
	Arg1 be a company base in Arg2	0.022

(*continued*)

Table 5. (*continued*)

Relation	Top 5 frequent patterns with prob.	
(Company, People)#1	*Arg1 work at Arg2*	*0.288*
	Arg1 be found by Arg2	*0.233*
	Arg1 to work for Arg2	*0.089*
	Arg1 be hire by Arg2	*0.056*
	Arg1 to work at Arg2	*0.031*

Table 5 also shows the top 1 frequent relation (represented using its top 5 patterns) of the above 4 argument type pairs. From Table 5 we can see that:

(1) Our method can group patterns which may implicitly express the same relation. For example, in Table 5 the pattern "*Arg1 be sold to Arg2*" can entail "*Arg1 be list a constituent of Arg2*", and the pattern "*Arg1 be headquarter in Arg2*" usually entails "*Arg1 establish in Arg2*".

(2) Some relations are hard to be distinguished from each other, because they are highly coupled in different documents. For example, in the *(company, people)* document, because the founder of a company will also work at that company, it will be hard to distinguish the *Be-Found-By* relation with the *Work-At* relation.

5   Related Work

In this section, we briefly review the related work of relation discovery. Start from the Message Understanding Conferences (MUC) (Grishman and Sundheim 1996), most relation extraction work focuses on supervised relation extraction methods, i.e., identifying and classifying relation instances within a document, given the annotated corpus and the target relation types. However, due to the large amount of manual engineering for corpus annotation and the large size of relations, recent research has focused on weakly supervised and self-supervised relation extraction, such as *DIPRE* (Brin 1999), *Snowball* (Agichtein and Gravano 2000), *KnowItAll* (Etzioni et al. 2004), *TextRunner* (Yates et al. 2007) and *NELL* (Carlson et al. 2010). The idea of these weakly supervised methods is to exploit the duality between relation instances and patterns, then a bootstrapping process can be constructed to iteratively extract new instances of the given relations.

In recent years, with the population of knowledge sharing web sites, a lot research efforts have been devoted to harvest machine-readable knowledge from Wikipedia, some projects include *Yago* (Suchanek et al. 2008), *DBpedia* (Auer et al. 2007) and *Kylin* (Wu and Weld 2007). The shortage of these projects is that they usually only harvest knowledge from the structures whose semantics is explicitly given, mostly the *Infoboxes* in Wikipedia. There were also some other research focus on building a relation extraction system using the distant supervision methods (Mintz et al. 2009), or organize the relation pattern using argument taxonomy hierarchy (Nakashole et al. 2012).

Some other work focuses on relation discovery from single domain corpus (Chen et al. 2011; Mohamed et al. 2011). The idea of these methods is to exploit the regularity in different syntactic levels, then identify the salient syntactic patterns in a domain as discovered relations.

6 Conclusions

This paper proposes a method which can discover a large collection of semantic relations from Wikipedia by exploiting the regularity and redundancy of semantic relations, and finally *14,299* relations, *105,661* patterns and *5,214,175* instances are discovered from Wikipedia. For future work, we want to exploit the argument type hierarchy in our method, so that the relations under lower level argument types can be inherited from their ancestor argument types. For example, the *Be-Married-To* relation of *(Actor, Actor)* can be inherited from *(People, People)*.

References

Agichtein, E., Gravano, L.: Snowball: extracting relations from large plain-text collections. In: Proceedings of the Fifth ACM Conference on Digital Libraries, pp. 85–94. ACM, New York (2000)

Amigo, E., Gonzalo, J., Artiles, J., Verdejo, F.: A comparison of extrinsic clustering evaluation metrics based on formal constraints. Ident. Common Mol. Subsequences **12**, 461–486 (2009)

Auer, S., Bizer, C., Kobilarov, G., Lehmann, J., Cyganiak, R., Ives, Z.: DBpedia: a nucleus for a web of open data. In: Aberer, K., Choi, K.-S., Noy, N., Allemang, D., Lee, K.-I., Nixon, L., Golbeck, J., Mika, P., Maynard, D., Mizoguchi, R., Schreiber, G., Cudré-Mauroux, P. (eds.) ASWC/ISWC -2007. LNCS, vol. 4825, pp. 722–735. Springer, Heidelberg (2007). doi:10.1007/978-3-540-76298-0_52

Baker, C.F., Charles, J.F., John, B.L.: The Berkeley framenet project. In: Proceedings of the 36th Annual Meeting of the Association for Computational Linguistics and 17th International Conference on Computational Linguistics, pp. 86–90. Association for Computational Linguistics, Stroudsburg (1998)

Banko, M., Cafarella, M.J., Soderland, S., Broadhead, M., Etzioni, O.: Open information extraction from the web. In: IJCAI, vol. 7, pp. 2670–2676 (2007)

Bunescu, R., Mooney, R.: A shortest path dependency kernel for relation extraction. In: Proceedings of the Conference on Human Language Technology and Empirical Methods in Natural Language Processing, pp. 724–731. Association for Computational Linguistics, Stroudsburg (2005)

Brin, S.: Extracting patterns and relations from the world wide web. In: International Workshop on the World Wide Web and Databases, pp. 172–183 (1999)

Carlson, A., Betteridge, J., et al.: Toward an architecture for never-ending language learning. In: Proceedings of the Conference on Artificial Intelligence (AAAI 2010), p. 3. AAAI Press, Palo Alto (2010)

Chan, Y.S., Roth, D.: Exploiting syntactico-semantic structures for relation extraction. In: Proceedings of the 49th Annual Meeting of the Association for Computational Linguistics: Human Language Technologies, pp. 551–560 (2011)

Chen, H., Benson, E., et al.: In-domain relation discovery with meta-constraints via posterior regularization. In: Proceedings of the 49th Annual Meeting of the Association for Computational Linguistics: Human Language Technologies, pp. 530–540. Association for Computational Linguistics, Stroudsburg (2011)

De Marneffe, M.C., Manning, C.D.: Stanford typed dependencies manual. Technical report, Stanford University, pp. 338–345 (2008)

Doddington, G., et al.: The automatic content extraction (ACE) program–tasks, data, and evaluation. In: Proceedings of LREC (2004)

Escobar, M.D., West, M.: Bayesian density estimation and inference using mixtures. J. Am. Stat. Assoc. **90**(430), 577–588 (1995)

Etzioni, O., Banko, M., et al.: Open information extraction from the web. Commun. ACM **51**, 68–74 (2008)

Etzioni, O., et al.: Web-scale information extraction in knowitall: (preliminary results). In: Proceedings of the 13th International Conference on World Wide Web, pp. 100–110. ACM, New York (2004)

Grishman, R., Sundheim, B.: Message understanding conference-6: a brief history. In: Proceedings of the 16th International Conference on Computational Linguistics, pp. 466–471 (1996)

Han, X., Sun, L.: An entity-topic model for entity linking. In: Proceedings of EMNLP-CoNLL, pp. 105–115. Association for Computational Linguistics, Stroudsburg (2012)

Li, P., Jiang, J., et al.: Generating templates of entity summaries with an entity-aspect model and pattern mining. In: Proceedings of ACL, pp. 640–649. Association for Computational Linguistics, Stroudsburg (2010)

Matuszek, C., Cabral, J., Witbrock, M., DeOliveira, J.: An introduction to the syntax and content of Cyc. In: Proceedings of the 2006 AAAI Spring Symposium on Formalizing and Compiling Background Knowledge and its Applications to Knowledge Representation and Question Answering, pp. 44–49. AAAI Press, Palo Alto (2006)

Miller, G.A.: WordNet: a lexical database for English. Commun. ACM **38**, 39–41 (1995)

Mintz, M., Bills, S., Snow, R., Jurafsky D.: Distant supervision for relation extraction without labeled data. In: Proceedings ACL-IJCNLP, pp. 1003—1011. Association for Computational Linguistics, Stroudsburg (2009)

Mohamed, T.P., Hruschka, J.E.R., et al.: Discovering relations between noun categories. In: Proceedings of EMNLP, pp. 1447–1455. Association for Computational Linguistics, Stroudsburg (2011)

Nakashole, N., Weikum, G., Suchanek, F.: PATTY: a taxonomy of relational patterns with semantic types. In: Proceedings of EMNLP, pp. 1135–1145 (2012)

Ponzetto, S.P., Navigli, R.: Large-scale taxonomy mapping for restructuring and integrating Wikipedia. In: Proceedings of the 21th IJCAI, pp. 2083–2088. AAAI Press, Palo Alto (2009)

Suchanek, F.M., Kasneci, G., et al.: Yago: a large ontology from wikipedia and wordnet. Web Semant.: Sci. Serv. Agents World Wide Web **6**, 203–217 (2008)

Teh, Y.W., Jordan, M.I., et al.: Hierarchical Dirichlet processes. J. Am. Stat. Assoc. **101**, 1566–1581 (2006)

Wang, C., Kalyanpur, A., et al.: Relation extraction and scoring in DeepQA. IBM J. Res. Dev. **56**, 9:1–9:12 (2012)

Wu, F., Weld, D.S.: Autonomously semantifying wikipedia. In: Proceedings of CIKM, pp. 41–50. ACM, New York (2007)

Yates, A., et al.: TextRunner: open information extraction on the web. In: Proceedings of HLT-NAACL, pp. 25–26. Association for Computational Linguistics, Stroudsburg (2007)

Biomedical Event Trigger Detection Based on Hybrid Methods Integrating Word Embeddings

Lishuang Li[✉], Meiyue Qin, and Degen Huang

School of Computer Science and Technology,
Dalian University of Technology, Dalian, China
{lilishuang314,qinmessiy}@163.com,
huangdg@dlut.edu.cn

Abstract. Trigger detection as the preceding task is of great importance in biomedical event extraction. By now, most of the state-of-the-art systems have been based on single classifiers, and the words encoded by one-hot are unable to represent the semantic information. In this paper, we utilize hybrid methods integrating word embeddings to get higher performance. In hybrid methods, first, multiple single classifiers are constructed based on rich manual features including dependency and syntactic parsed results. Then multiple predicting results are integrated by set operation, voting and stacking method. Hybrid methods can take advantage of the difference among classifiers and make up for their deficiencies and thus improve performance. Word embeddings are learnt from large scale unlabeled texts and integrated as unsupervised features into other rich features based on dependency parse graphs, and thus a lot of semantic information can be represented. Experimental results show our method outperforms the state-of-the-art systems.

Keywords: Trigger detection · Word embeddings · Hybrid methods · Rich features

1 Introduction

With the development of the Internet, a vast and ever-expanding body of natural language text is becoming increasingly difficult to leverage. This is particularly true in the domain of life science, where biomedical articles are increasing exponentially. We need to automatically extract interested and structured information from biomedical text, which is known as biomedical text mining. In the past, the focus in the field of biomedical text mining was named entity recognition (NER). In recent years, the focus has shifted to relation extraction, especially complex relation extraction which is more difficult than simple binary relation extraction. Biomedical event is one type of complex relation. Trigger, argument and the event type need to be detected when extracting an event. Event extraction systems consist of at least two parts: trigger detection and argument detection, while trigger detection is the preceding task. Thus, trigger detection is of great importance in biomedical event extraction.

© Springer Nature Singapore Pte Ltd. 2016
H. Chen et al. (Eds.): CCKS 2016, CCIS 650, pp. 67–79, 2016.
DOI: 10.1007/978-981-10-3168-7_7

Trigger detection aims to detect a span of text that triggers an event. The methods for trigger detection fall into four categories: dictionary-based, rule-based, statistical machine learning and combined methods in which the statistical machine learning method is dominant. Trigger detection is regarded as multiclass classification task in most of the state-of-art event extraction systems. Björne et al. extracted rich manual features including token features, frequency features and dependency chains and so on [1]. They adopted these features and multiclass classification tool $SVM^{multiclass}$ to detect triggers. Their event extraction system achieved the best performance using this trigger detection. Martinez and Baldwin regarded trigger detection as word sense disambiguation (WSD) problem and found that WSD outperformed sequential tagging and could improve the performance of sequential tagging methods [2]. They achieved 60.1% F-score on the set of BioNLP'09. Zhang et al. efficiently mapped the dependency graph of a candidate sentence into semantic/syntactic features, and used these semantic/syntactic features to detect bio-event triggers from the biomedical literature [3]. Their method achieved an F-score of 65.84% on the set of BioNLP'09. Trigger detection was viewed as sequence labeling task by reference [4]. They designed elaborate features, such as the frequency of named entity in sliding window, dependency path and adopt Conditional Random Field (CRF) to extract triggers with feature template. The F-score achieved 67.0% on the set of BioNLP'09. Wang et al. proposed a method based on the deep syntactic analysis [5]. They adopted deep syntactic information to detect triggers and arguments with LibSVM. The results from arguments detection were integrated into trigger detection. They achieved 68.8 and 67.3% F-scores on BioNLP'09 and BioNLP'11 respectively.

The previous works were mostly based on single models, Domingos pointed out one model was not sufficient [6]. On one task, many models can be constructed and their results can be combined based on different techniques. Ensemble techniques include set operation, voting and stacking, etc. Li et al. discussed the three techniques on NER task which was regarded as a sequence labeling problem [7]. Due to the relearning process in stacking, the stacking technique outperformed the other two which directly operated the predict results. Similar to NER, the statistical machine learning methods for trigger detection can be integrated under the construction of single models. In this work, we construct four different models based on two SVM models trained separately using one vs. one and one vs. rest multiclass extension methods, Passive aggressive online algorithm (PA) [8] and Random Forest (RF) [9]. And then the results from four models are integrated with different ensemble techniques.

On the other hand, the way to digitalize features in previous works was one-hot encoding. The main problem of this method is that it is unable to represent the semantic information. Recently, word embeddings, a vector related with a word, are used in several NLP problems, such as named entity recognition (NER), chunking, and make a contribution to the improvement. Tang et al. explored the effect of word embeddings on biomedical NER [10]. Turian et al. discussed its effect on several tasks, including NER and chunking [11]. In this work, we utilize hybrid methods integrating word embeddings to predict trigger in biomedical event. Experimental results show our method outperforms the state-of-the-art systems.

The remaining part of this paper is organized as follows: Our proposed method is described in Sect. 2. Experimental results and analysis are illustrated in Sect. 3. Comparisons are given in Sect. 4. Finally, discussion and conclusions are shown in Sects. 5 and 6 respectively.

2 Our Methods

2.1 Word Embeddings

A distributed representation, also known as word embeddings, is dense, low dimensional, and real-valued. Word embeddings are typically induced using neural language models, which uses neural networks as the underlying predictive model. There are several word embeddings, such as Collobert and Weston embeddings (C&W) [12], HLBL embeddings [13] and Word2Vec [14, 15].

Considered the time and hardware requirements in different distributed representation methods, Word2Vec was adopted in our work. Word2Vec supplies two models: CBOW and Skip-gram. The Skip-gram model extended on n-gram model is used. It aims to optimize the classification of a word based on other words in the same sentence within a certain range before and after the current word. This tool can generate a dense, low-dimensional, and real-valued vector, which may capture the syntactic and semantic information in each dimension. This information cannot be obtained from words encoded by one-hot.

2.2 Features Extraction

In this work, five kinds of features are mainly used, token, frequency, dependency chains, shortest path and word embeddings. The dependency paths parsed by McClosky-Charniak parser [16] and Enju parser [17] are added into the features. Compare to the previous researches in BioNLP, our system extracts more features, which have greatly improved the performance. The features we employ are:

Token features include current token text, POS, stem, binary tests for presence of uppercase, digital or special characters, bigrams and trigrams of the token. Dependency context is of great importance for trigger detection, so we extract token features of candidate triggers in dependency context and linear context besides candidate triggers themselves.

Frequency features are defined as the number of named entities in the current sentence and the context of a candidate trigger, and the frequency of words in bag-of-words. It is obvious that the more entities in a sentence there are, the more likely triggers exist in the current sentence.

Dependency chains up to depth of three are constructed. When the window size is not large enough, the important information related with candidate triggers may not be considered. Therefore, dependency information is added.

Token features of nodes in dependency chains include POS of the token, the token and whether the node is protein or not. These features are added with position information (the distance from proteins) in dependency chains.

Dependency types in dependency chains are also added with position information, sequence of dependency type and direction.

Shortest path includes *n*-grams ($n = 2, 3, 4$) of the edges in the shortest dependency path between candidate triggers and the nearest protein, and the combinations of the entity types in the shortest path. For more details, please refer to [18].

Word embeddings involve the vectors of the current token. The dimension of the vectors is decided by experiments.

2.3   Divergent Classifiers

In our experiment, we utilize three different toolkits and adopt different training algorithms to construct four classifiers.

- PA [8]: follows the maximum edge theory and has good generalization ability like SVM [19].
- SVM1vs1 and SVM1vsrest: two SVM models trained by one vs. one multi-class extension method and one vs. rest multi-class extension method.
- RF [9]: a combination of tree predictors, and after a large number of trees are generated, they vote for the most popular class.

2.4   Hybrid Methods

Our system uses three different ensemble methods which are set operations, voting methods and stacking method to combine the four single classifiers' results. Firstly, the set operations and voting methods which do not need retraining process are adopted to combine the classification results from the four models. They both cost less time than the stacking method because the latter needs retraining. For example, the stacking method with *n*-fold cross evaluation on the training corpus costs much more training time than the combining methods with no retraining. The three hybrid methods are presented in detail in following sections.

2.5   Union and Intersection Operation

According to the union operation, both classification results from two classifiers are classified as the correct results. Obviously this method will make the recall improved but make the precision decreased compared to each single classifier. On the contrary, the intersection of two classifiers will only take the common results as the correct results, which will make the precision improved but the recall decreased. In order to make a trade-off between recall and precision, we perform union or intersection operations on the results from different models depending on precision and recall of different models.

2.6   Voting

The majority voting method assumes that triggers are correctly predicted by most individual systems while different systems cannot get consistent results. The pseudo code of the voting method used in this paper is described in below:

```
Input: predicting result of single classifiers for one
trigger instance;

Output: predicting type of trigger;

Voting:

  Initial: set result_voting to 1 by default and
  elements of array count to 0,
  count[1],count[2],…,count[10] represent the number of
  vote for each class, respectively;

  Calculate count[1],count[2],…,count[10];

  max_value, index = the max value in array count and
  the index of max value in array respectively;

  if max_value == 1:

    result_voting = the highest prediction result of
    single classifiers;

  else:

    result_voting = index;

return result_ voting.
```

2.7   Stacking Method

Most stacking methods adopt the two-layer framework. The training process is separated into two steps and is described as follows:

- Step 1: n-fold cross validation is adopted on the single classifier of the layer-0. Given a data set $D = \{(x_1, y_1), \ldots, (x_m, y_m)\}$ and k different learning algorithms, we split the data set D into n almost equal parts; At each training and testing process, choosing one part as testing corpus and the other $n-1$ parts as training corpus; For this part of testing corpus, we get k different classification results from the k classifiers. After n times training and testing like this, we get k different results on entire data set D; then we combine the k results and the manually annotated results of D, and then get a new training set D_1 for the layer-1.
- Step 2: At this step D_1 is utilized as the training corpus to construct a classifier model based on a learning algorithm and its testing results on the testing corpus are the final results.

The four classifiers described in Sect. 2.3 are used as the base classifiers at layer-0, and RF is chosen as the classifier at layer-1 because the framework in terms of strength of the individual predictors and their correlations gives insight into the ability of the random forest to predict.

In the training process, we use 5-fold cross validation to get the predicting results of the four kinds of classifiers on BioNLP'09 and BioNLP'11 training sets respectively. Then we regard the four results as feature vectors to construct a new training set for the classifier at layer-1. Another work we have to do at layer-0 is constructing four classifiers based on the whole training corpus and predicting the classification results on BioNLP'09 and BioNLP'11 development sets based on them respectively. In the same way we combine the four results of single classifiers and get the new testing corpus for the classifier at layer-1. The two-layer stacking frame is shown in Fig. 1.

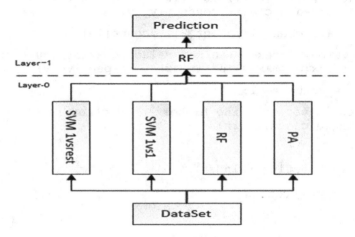

Fig. 1. Two-layer stacking architecture of hybrid method.

3 Experiments and Results

3.1 Corpus and Evaluation

All experiments are conducted on the corpora supplied by BioNLP'09 [20] and BioNLP'11 [21]. And the parameters are optimized by using 5-fold cross evaluation on training set. The evaluation criterion P(recision)/R(ecall)/F(-score) is adopted, which is defined as formula (1), where *TP*, *FP* and *FN* are short for True Positives, False Positives and False Negatives respectively.

$$P = \frac{TP}{TP+FP}, \quad R = \frac{TP}{TP+FN}, \quad F = \frac{2*P*R}{P+R}. \tag{1}$$

3.2 Results of Trigger Detection Integrating Word Embeddings Based on PA

To illustrate the impact of word embeddings on trigger detection, we choose PA without word embeddings as baseline. Five groups of experiments are conducted on the development set of BioNLP'09 with different dimensions of the word vectors. The dimension of the vectors is set to 50, 100, 200 and 400 respectively to compare the influence of word embeddings on trigger prediction. The results are shown in Table 1, and our baseline is using all features except word embeddings. BaselineWE50, BaselineWE100, BaselineWE200 and BaselineWE400 mean the dimensions of word embeddings are 50, 100, 200 and 400 respectively when word embeddings are integrated. The type with the highest score is the final result.

From Table 1, we can see all the F-scores using word embeddings are improved compared with Baseline. The F-score improves with the increase of dimension on trigger prediction. The F-scores are improved by $0.39 \sim 1.24\%$ with the variance of the dimension of the vectors, which illustrates that the syntactic and semantic information carried by word embeddings has significantly increased the performance.

Table 1. The results with different dimensions of the word vectors on trigger prediction.

Features	P	R	F
Baseline	72.19%	71.33%	71.76%
BaselineWE50	73.93%	70.45%	72.15%
BaselineWE100	74.44%	71.41%	72.89%
BaselineWE200	74.17%	71.81%	72.97%
BaselineWE400	74.58%	71.49%	**73.00%**

3.3 Results Based on Four Single Classifiers

Tables 2 and 3 shows the results from the four single classifiers in shared task BioNLP'09 and BioNLP'11 development sets respectively. We can see that the PA model constantly outperforms the other models. Although RF gets a higher Precision of 79.57%, the lowest Recall leads to the lowest F-score (63.99%).

Table 2. Results based on four single classifiers on BioNLP'09.

Model	P	R	F
PA	74.58%	**71.49%**	**73.00%**
SVM 1vs1	74.30%	69.73%	71.94%
SVM 1vsrest	**80.02%**	64.30%	71.30%
RF	79.57%	53.51%	63.99%

Table 3. Results based on four single classifiers on BioNLP'11.

Model	P	R	F
PA	74.57%	**72.74%**	**73.64%**
SVM 1vsrest	**81.35%**	64.61%	72.02%
SVM 1vs1	73.39%	67.77%	70.47%
RF	78.06%	56.85%	65.79%

3.4   Results Based on Union and Intersection Operation Method

The performance combining the results of the different models using the simple set operations is shown in Table 4. We conduct the union operation (denoted by the symbol " ∪ ") on the results of SVM 1vsrest, SVM1vs1 and PA as they are all based on maximum-margin theory. It can be seen that Recall increases by 3.19% (74.68 vs 71.49%) compared with PA which achieves the highest Recall among the single classifiers. We also intersect (denoted by the symbol " ∩ ") them to improve Precision 3.33% (83.35% vs 80.02) higher than SVM1vsrest which gets the best Precision. On the basis of the above set operation, we try to use RF (not based on maximum-margin theory) to improve the performance. However, due to the poor performance of RF, it decreases the F-score (shown as the third and fourth row in Table 4). From Table 4, we can see that SVM1vsrest ∪ SVM1vs1 ∩ PA can get the best F-score of 73.38% which is 0.38% higher than PA (73%).

Table 4. Results using simple set operation methods on BioNLP'09.

Method	P	R	F
PA ∪ SVM 1vs1 ∪ SVM 1vsrest	70.25%	**74.68%**	72.40%
PA ∩ SVM 1vs1 ∩ SVM 1vsrest	83.35%	60.38%	70.03%
PA ∪ SVM 1vs1 ∪ SVM 1vsrest ∪ RF	68.90%	75.00%	71.82%
PA ∩ SVM 1vs1 ∩ SVM 1vsrest ∩ RF	**87.39%**	49.28%	63.02%
SVM_U:SVM 1vsrest ∪ SVM 1vs1	73.03%	72.44%	72.73%
SVM_U ∩ PA	78.45%	68.93%	**73.38%**
SVM_I:SVM 1vsrest ∩ SVM 1vs1	82.45%	61.18%	70.24%
SVM_I ∪ PA	74.22%	72.20%	73.20%

3.5   Results Based on Voting Method

Some experiments are conducted to investigate the effectiveness of the voting algorithm. The results are shown in Table 5. The voting method (PA + SVM1vs1 + SVM1vsrest) gets a better F-score (73.41%) than the simple set operations method. The reason may be that it's easy to reach agreement on the same instance with similar classifications, thus the voting results are more reliable. From the Table 5, we can also find that RF as one member of voting groups may decrease the final result because of its poor performance.

Table 5. Results based on voting method on BioNLP'09.

Method	P	R	F
PA + SVM1vs1 + SVM1vsrest	77.41%	**69.80%**	**73.41%**
RF + SVM1vs1 + SVM1vsrest	80.65%	63.90%	71.30%
PA + RF + SVM1vs1	77.38%	68.85%	72.87%
PA + RF + SVM1vsrest	80.44%	64.70%	71.71%
PA + SVM1vs1 + SVM1vsrest + RF	**81.71%**	63.18%	71.26%

3.6 Results Based on Stacking Method

The following three groups of experiments are conducted in the stacking method: (1) Choose two classifiers which get the best Recall and Precision as base classifiers at layer-0 and the stacking results are regarded as our baselines, denoted by baseline 1 and baseline 2 respectively. (2) Add different classifiers to the baselines. From Tables 6 and 7 it can be seen that after adding a different classifier, all of F-scores are improved than both baselines respectively. We can also find that adding RF can get the better performance though its performance is poor. Therefore, the diversity among different classifiers plays an important role in stacking method. (3) Use all four classification results as base classifiers at layer-0. We can find that the F-score is under baseline1, which means that more classifiers may not achieve better performance.

From Table 6, we can find that the group of PA + SVM1vs1 + SVM1vsrest can get the best performance (73.79% F-score) on BioNLP'09 which is 0.79% higher than PA which achieves the best F-score (73%) among single classifiers. The same stacking experiments are executed in task BioNLP'11 (shown as Table 7), and we can get a similar conclusion. Compared to the single classifier's best F-score (73.64%), the stacking method improve the F-score by 0.61% on BioNLP'11.

Table 6. Results based on two-layer stacking method on BioNLP'09.

Layer-0 method	P	R	F
PA + SVM1vs1 (baseline 1)	76.12%	**70.29%**	73.09%
RF + SVM1vsrest (baseline 2)	**80.20%**	64.38%	71.42%
PA + SVM1vs1 + SVM 1vsrest	77.96%	70.05%	**73.79%**
PA + SVM1vs1 + RF	77.46%	69.73%	73.39%
RF + SVM1vsrest + PA	78.88%	67.73%	72.88%
RF + SVM1vsrest + SVM 1vs1	79.18%	66.53%	72.31%
PA + SVM1vsrest + SVM1vs1 + RF	78.56%	68.21%	73.02%

4 Comparisons

4.1 Comparisons of Performance of Different Methods

The comparison among the results of the union and intersection operation methods, voting algorithm, two-layer stacking method and the single classifier PA is shown in Table 8. Here we regard the result of PA as a baseline because of its best performance

Table 7. Results based on the two-layer stacking method on BioNLP'11.

Layer-0 method	P	R	F
PA + SVM1vs1 (baseline 1)	75.58%	72.60%	74.06%
RF + SVM1vsrest (baseline 2)	**81.02%**	65.03%	72.15%
PA + SVM1vs1 + SVM 1vsrest	75.81%	**72.64%**	74.19%
PA + SVM1vs1 + RF	76.13%	72.46%	**74.25%**
RF + SVM1vsrest + PA	78.31%	69.25%	73.50%
RF + SVM1vsrest + SVM 1vs1	80.86%	65.35%	72.28%
PA + SVM1vsrest + SVM1vs1 + RF	77.64%	69.35%	73.26%

Table 8. Comparisons of performance on different methods on BioNLP'09.

Method	P	R	F
PA (baseline)	74.58%	71.49%	73%
Union and intersection method	78.45%	68.93%	73.38%
Voting (three classifiers)	77.41%	69.80%	73.41%
Two-layer stacking algorithm	77.96%	70.05%	**73.79%**

among four single classifiers. From Table 8 we can see that all hybrid methods yield better results than each single classifier. Compared with PA, the three different ensemble methods all improve the precision but decrease the recall in task BioNLP'09. Furthermore, the two-layer stacking method achieves better performance than the other two hybrid methods.

4.2 Comparisons with Other Work

Finally, we make comparisons between our systems and some related work in Table 9. We achieve the best performance on BioNLP'09 and BioNLP'11 development sets. The F-scores are higher than the current best system [5] by 4.99 and 6.95% respectively.

Wang et al. proposed a trigger extraction method based on the deep syntactic analysis [5]. Deep syntactic information was used for argument detection, and then the result was merged into the trigger extraction phase. This method achieved 68.8 and 67.3% F-scores

Table 9. Comparisons between our system and some related work.

System	Task	P	R	F
Ours	BioNLP'09	77.96%	70.05%	**73.79%**
	BioNLP'11	76.13%	72.46%	**74.25%**
Wang et al.'s [5]	BioNLP'09	75.30%	64.00%	68.80%
	BioNLP'11	69.50%	56.90%	67.30%
Martinez and Baldwin's [2]	BioNLP'09	70.20%	52.60%	60.10%
Majumder's [4]	BioNLP'09	69.96%	64.28%	67.00%
Zhang et al.'s [3]	BioNLP'09	79.83%	56.02%	65.84%

on BioNLP'09 and BioNLP'11 respectively. Martinez et al. regarded trigger classification as a word sense disambiguation (WSD) problem [2]. In the task of BioNLP'09, the F-score reached 60.1%. Majumder took trigger classification as a sequential tagging task and extracted rich features such as frequency of named-entities in sliding window, POS of word, whether protein or others and name of nearest protein etc. [4]. They used CRF tool to tag sequences and achieved an F-score of 67.0% on BioNLP'09. Zhang et al. used the hash operation to iteratively compute the dependency graph and mapped the dependency graph into neighborhood hash features [3]. Then they combined other basic features, bag-of-words features, frequency features and token features based on SVM. Finally, their approach achieved an F-score of 65.84% on BioNLP'09.

The main difference between our method and the other four methods exists in three aspects: (1) The rich features are the solid foundation, such as token features, syntactic and dependency features, the shortest path. (2) Word embeddings, which can learn much deeper syntactic and semantic information from the large set of out-of-domain data obtained through unsupervised learning, lead to the vectors of words with common semantics are close to each other, and thus improve trigger detection. (3) Hybrid methods: multiple classification results are combined to further improve the performance.

5   Discussion

The three ensemble methods give better performance than every single model. The main reason is that the hybrid methods can exploit the diversity or consistency among different classifiers to make a final decision on the basis of single models. For instance, the trigger "transfection" is classified as Regulation by PA, on the contrary, all of the other classifiers categorized it as Positive regulation". After voting it is marked as "Positive regulation" which is consistent with the correct result.

Among all the three hybrid methods in our paper (set operation, voting and stacking), the stacking method performs best owing to its capability of relearning from the original learning at layer-0. In the relearning process for RF, after a large number of trees are generated, they vote for the most popular class. For example, "Overexpression", which is categorized as Regulation by voting according to most classifiers' results, can be marked correctly as "Gene expression" by the stacking method through the relearning process.

Word embeddings play an important role which implies a lot of useful information, including syntactic and semantic. For example, for the two words, "diminished" and "reduced", they have little common features directly in morphology, but the similarity between their word embeddings measured by cosine similarity is up to 0.897. By using word embedding, the performance on trigger prediction is improved.

6   Conclusion

The proposed method improves the performance of trigger detection, outperforming most of published works. First, rich features are the solid foundation. Second, word embeddings play an important role. Finally, the hybrid methods make full use of the advantages of different classifiers by combining their results to get a higher performance. By integrating the rich features and word embeddings into hybrid method, our system outperforms the state-of-the-art systems.

Acknowledgments. The authors gratefully acknowledge the financial support provided by the National Natural Science Foundation of China under No. 61672126, 61173101.

References

1. Björne, J., Heimonen, J., Ginter, F., Airola, A., Pahikkala, T., Salakoski, T.: Extracting complex biological events with rich graph-based feature sets. In: Proceedings of Workshop on Current Trends in Biomedical Natural Language Processing: Shared Task, pp. 10–18. ACL, Boulder, Colorado (2009)
2. Martinez, D., Baldwin, T.: Word sense disambiguation for event trigger word detection in biomedicine. BMC Bioinform. **12**(Suppl. 2), S4 (2011)
3. Zhang, Y., Lin, H., Yang, Z., Wang, J., Li, Y.: Biomolecular event trigger detection using neighborhood hash features. J. Theoret. Biol. **318**, 22–28 (2013)
4. Majumder, A.: Multiple features based approach to extract bio-molecular event triggers using conditional random field. Int. J. Intell. Syst. Appl. **4**(12), 41–47 (2012)
5. Wang, J., Wu, Y., Lin, H., Yang, Z.: Biological event trigger extraction based on deep parsing. Comput. Eng. **39**, 25–30 (2013)
6. Domingos, P.: A few useful things to know about machine learning. Commun. ACM **55**(10), 78–87 (2012)
7. Li, L., Fan, W., Huang, D., Dang, Y., Sun, J.: Boosting performance of gene mention tagging system by hybrid methods. J. Biomed. Inf. **45**(1), 156–164 (2012)
8. Crammer, K., Dekel, O., Keshet, J., Shalev-Shwartz, S., Singer, Y.: Online passive-aggressive algorithms. J. Mach. Learn. Res. **7**, 551–585 (2006)
9. Breiman, L.: Random forests. Mach. Learn. **45**, 5–32 (2001)
10. Tang, B., Cao, H., Wang, X., Chen, Q., Xu, H.: Evaluating word representation features in biomedical named entity recognition tasks. BioMed. Res. Int. **2014**, Article ID 240403, 1–6 (2014). Hindawi Publishing Corporation
11. Turian, J., Ratinov, L., Bengio, Y.: Word representations: a simple and general method for semi-supervised learning. In: Proceedings of 48th Annual Meeting of the Association for Computational Linguistics, Uppsala, Sweden, pp. 384–394 (2010)
12. Collobert, R., Weston, J., Bottou, L., Karlen, M., Kavukcuoglu, K., Kuksa, P., Collins, M.: Natural language processing (almost) from scratch. J. Mach. Learn. Res. **12**, 2493–2537 (2011)
13. Mnih, A., Hinton, G.: A scalable hierarchical distributed language model. In: NIPS, pp. 1081–1088 (2008)
14. Mikolov, T., Sutskever, I., Chen, K., Corrado, G., Dean, J.: Distributed representations of words and phrases and their compositionality. Adv. Neural Inf. Process. Syst. **26**, 3111–3119 (2013)
15. Mikolov, T., Yih, W.T., Zweig, G.: Linguistic regularities in continuous space word representations. In: Proceedings of NAACL-HLT, Atlanta, Georgia, pp. 746–751 (2013)
16. McClosky, D., Charniak, E.: Self-training for biomedical parsing. In: Proceedings of 46th Annual Meeting of the Association for Computational Linguistics on Human Language Technologies, Columbus, Ohio, pp. 101–104 (2008)

17. Miyao, Y., Sagae, K., Saetre, R., Matsuzaki, T., Tsujii, J.: Evaluating contributions of natural language parsers to protein–protein interaction extraction. Bioinformatics **25**(3), 394–400 (2009)
18. Miwa, M., Saetre, R., Kim, J.D., Tsujii, J.: Event extraction with complex event classification using rich features. J. Bioinform. Comput. Biol. **8**(1), 131–146 (2010). doi:10.1142/S0219720010004586
19. Vapnik, V.N.: The Nature of Statistical Learning Theory. Springer, Berlin (1995)
20. Kim, J.D., Ohta, T., Pyysalo, S., Kano, Y., Tsujii, J.: Overview of BioNLP'09 shared task on event extraction. In: Proceedings of Workshop on BioNLP: Shared Task, Boulder, Colorado, pp. 1–9 (2009)
21. Kim, J.D., Pyysalo, S., Ohta, T., Bossy, R., Nguyen, N., Tsujii, J.: Overview of BioNLP shared task 2011. In: Proceedings of BioNLP Shared Task 2011 Workshop, pp. 1–6. Association for Computational Linguistics, Portland (2011)

GRU-RNN Based Question Answering Over Knowledge Base

Shini Chen, Jianfeng Wen, and Richong Zhang[✉]

State Key Laboratory of Software Development Environment,
School of Computer Science and Engineering,
Beihang University, Beijing, China
zhangrc@act.buaa.edu.cn

Abstract. Building system that could answer questions in natural language is one of the most important natural language processing applications. Recently, the raise of large-scale open-domain knowledge base provides a new possible approach. Some existing systems conduct question-answering relaying on hand-craft features and rules, other work try to extract features by popular neural networks. In this paper, we adopt recurrent neural network to understand questions and find out the corresponding answer entities from knowledge bases based on word embedding and knowledge bases embedding. Question-answer pairs are used to train our multi-step system. We evaluate our system on FREEBASE and WEBQUESTIONS. The experimental results show that our system achieves comparable performance compared with baseline method with a more straightforward structure.

1 Introduction

Recently, some structured knowledge bases have been published, such as Freebase [6], DBpedia [1], YAGO [17]. The vertices and edges in these graphs with different labels represent entities and relations in real world. The availability of knowledge bases makes it possible to discover relational knowledge from clean and structured data storage. Especially when we using human language to query the knowledge base, mapping the question text with the stored knowledge is a great challenge.

To map the search desire to the triples in the knowledge base, most of the existing studies [7,11] focus on understanding the question and finding the matching entities and relations in the knowledge base. One of the characteristic features of the knowledge base or knowledge graph is that there exists a fix number of relations. However, the user provided questions may vary significantly. The key issue for successfully locating the correct answers for a question is to discover the hidden links between the questions' syntactical structures and the relations. In practice, the questions' syntactical structures usually follow some specific patterns. In this study, we advocate that the latent semantic matching between the question and the knowledge triple provides an opportunity to model

H. Chen et al. (Eds.): CCKS 2016, CCIS 650, pp. 80–91, 2016.
DOI: 10.1007/978-981-10-3168-7_8

the hidden relation between question patterns and the relations in the knowledge base. Specifically, we propose three steps for solving the question answering problem over knowledge base. For the relation identification from the question text and question mapping on the knowledge base, further than existing deep learning method which simply put corpus into deep neural networks, we design a two-column GRU-based RNN for characterizing the latent semantics between question text and the knowledge triples. Empirical studies on the commonly-used WEBQUESTIONS for question answering task evaluation confirms the effectiveness of our proposed model for both relation identification and question and answer mapping.

The remainder of this paper is organized as follows. Section 2 introduces the related studies for solving the problem of question answering over the knowledge base. Section 3 delivers the procedures and the two-column GRU-based RNN model. Section 4 describes the training and inference of the proposed model. Section 5 presents experimental evaluation of our framework. Finally, Sect. 5 will discuss and conclude the paper.

2   Related Work

The state-of-art method in knowledge based QA can be divided into two mainstreams, namely, semantic parsing based and information retrieval based.

Semantic parsing based method focus on learning semantic parsers which parse natural language question into logical form and query knowledge base to lookup answers. In [4], authors propose an approach that generates query candidates by recursively generating logical form with a mapping of phrases to knowledge base predicates and a small set of composition rules, and rank query candidates by log-linear model. In [5], a set of candidate logical form is generated and then a paraphrase model is introduced to choose the realization that best paraphrases the input question, and the corresponding logical form is produced. While early approaches heavily relied on manually annotated logical form and high-quality lexicons to train semantic parser, recent work has focused on training semantic parser only using question answer pair. In [3], the proposed approach translates a given natural language question to the matched SPARQL query and use learning-to-rank techniques to learn pair-wise comparison of query candidate. [19] formulates semantic parsing as a staged search problem, mapping natural language question into a query graph which resembles subgraph of knowledge graph.

Information retrieval based method first retrieves a large set of candidate answers from knowledge base, and then rank them by fine-grained extracted features from question and answer. In [18], authors propose a model for directly learning the pattern of question answer pair. Firstly, question dependency parse is converted to candidates topic graphs by rules, then the relations and properties in topic graph are fed into a logistic regression model as features to classify correctness of questions candidate answer. Recently, many studies embed questions and knowledge graph entries in a low-dimensional vector spaces and retrieve the answers by computing similarities in learned embedding space. For example, [7] combines the embeddings of words in question as its representation, and encodes answer by summing embeddings of entities and relations that appear in question-answer path and surrounding subgraph. Then, the score of a question-answer pair is given by dot production of question embedding and answer embedding. In addition, [11] uses multi-column convolutional neural network to generate three question aspects, and ranks candidate by considering answer type, answer path and answer context. In [8], authors conduct QA process under the embedding based Memory Networks framework. In [12], the sequence translation framework are exploited to feed question characters into encoding LSTM and to obtain the knowledge base triples from an attention-based decoding LSTM.

In general, these existing studies focus on transformation from the question words to a knowledge base query. However, the sequential patterns of question word is ignored for building the QA model. In practice, this sequential information is important for the question understanding. In this study, we exploit the GRU-RNN model, which can characterize the sequential patterns of input question text, identify the question pattern, and match the question and the knowledge base triples.

3   Model

3.1   Problem Definition

In general, three aspects are considered for the question answering problem over a knowledge graph. The first aspect is to identify the entities that have appeared in the question. It focuses on identifying the words in the question that may be translated to the entities in the knowledge graph; the second aspect is to discover the relation mentioned in the question sentence; and the third aspect is to understand the semantics of the entities and relations and match them with the existing entities and relations or paths in the knowledge graph and then to rank the matched triples or paths. In this study, we propose three main modules for solving the above mentioned aspects.

3.2   Entity Matching Module

The question of how exactly the topic entity is identified has been discussed by many research, e.g. [14,15]. In this study, we exploit the solutions provided by [3] to identify the entity in the question. The match between the question words and the knowledge entities could be literal or via an alias of the entity name.

We first POS-tag a question by the Stanford tagger [16], and apply some simple rules to filter subsequences (n-grams) of question to get candidate entity word set. The rules are: (1) a subsequence containing only single word must tagged NN (noun) (2) consecutive words tagged NNP (proper noun) cannot be split into two subsequences. The filtered subsequence set S is used to retrieve a list of entities from knowledge base, whose name or alias is literally similar to a candidate in S. We use dictionary provided by [3], which contains mappings from name or alias to Freebase entities with matching scores. We set a threshold to limit the number of candidate entities.

3.3   Relation Identification Module

The most important step for retrieving the answer from a knowledge base is to identity the question semantics and to locate the corresponding knowledge graph relations. In practice, the questions' syntactical structure usually follow some patterns. If we remove the entity word from the question sentence, the remaining sequence of words can somehow represent the **question pattern**, or the semantic pattern of a question sentence. The relation in **FREEBASE** is organized as the format of *relation_field.excepted_subject_type.relation_name*, which can be considered as a sequence of sub-relation labels. For instance, *people.deceased_person.cause_of_death* can be considered as a sequences of people, deceased_person, cause_of_death.

The relation identification problem now is translated into modeling the semantic similarity between the **question pattern** and the knowledge base relation. For example, on one hand, question *what did george orwell died of* and question *what was jesse james killed with* should be mapped into the same knowledge base relation *people.deceased_person.cause_of_death*. On the other hand, the same knowledge base relation *people.deceased_person.cause_of_deat* may correspond to question patterns *what did _ died of* and *what was _ killed with*.

To discover the semantic relation between **question pattern** and the knowledge base relation (pattern-relation pair), we build a two-column GRU-based RNN, which is displayed in Fig. 1. We will introduce this model in the following subsections.

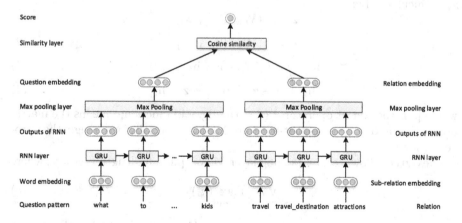

Fig. 1. The two-columns GRU-based RNN model

3.4   Question-Triple Matching Module

As the final goal of the QA task is to find the answers from the knowledge base, the relation between the question and the triple in the knowledge graph is to be determined. To model this relation, we translate the triple into a sequence of words by concatenating the subject, the relation and the object of a triple together. For example, the triple m.02khkd *people.deceased_person.cause_of_death* m.012hw is translated as *jesse james people deceased person cause of death assassination*. Here, *jesse james* and *assassination* are standard name of entity m.02khkd and m.012hw respectly. We use the same two-column GRU-based RNN, as shown in Fig. 1 to characterize the question-triple relations.

3.5   GRU-Based RNN for QA

Figure 1 shows the learning architecture for the relation identification and answer matching system. This architecture consists of two-column independent recurrent neural network (RNN) with Gated Recurrent Unit (GRU) cell [10]. This two-column GRU-RNN is used to character the similarity between two input sequences. Each GRU RNN layer takes a sequence of vectors as input, and produces a output vector for each input vector. In our system, the input are sequence of words or sub-relation labels, so we apply a lookup layer to transform them into vectors. For better understanding the latent semantics underlining the word sequences, we use the embedding as input to the GRU cell.

The lookup layer transforms every input word w_i of q or sub-relation label rl_i of the sub-relation label sequence into a input embedding vector $\mathbf{x_i} = \mathbf{W}\phi_i$, where ϕ_i is the one-hot vector representing w_i or rl_i, $\mathbf{W} \in \mathbb{R}^{k \times |Elements|}$ is the matrix of word embedding, k is the input vector dimension, $|Elements|$ is the total number of words and sub-relation labels. Output vector h_i for input vector x_i is calculated by:

$$\mathbf{z}_i = \sigma(\mathbf{W_z} \cdot [\mathbf{h}_{i-1}, \mathbf{x}_i]) \tag{1}$$

$$\mathbf{r}_i = \sigma(\mathbf{W_r} \cdot [\mathbf{h}_{i-1}, \mathbf{x}_i]) \tag{2}$$

$$\tilde{\mathbf{h}}_i = tanh(\mathbf{W_h} \cdot [\mathbf{r}_i * \mathbf{h}_{i-1}, \mathbf{x}_i]) \tag{3}$$

$$\mathbf{h}_i = (1 - \mathbf{z}_i) * \mathbf{h}_{i-1} + \mathbf{z}_i * \tilde{\mathbf{h}}_i \tag{4}$$

where [.] is the concat operator; $\sigma(\cdot)$ is the sigmoid function; $\cdot$ means the matrix production and $*$ is a element-wise production. Specially, we assign the zero-vector for h_0.

Next, the following max-pooling layer will output a fix-length vector $\mathbf{v}$, where

$$\mathbf{v} = \max_{i=1,\dots,n} (h_i) \tag{5}$$

the $\max(\cdot)$ is the element-wise operator over $\{h_i\}$ and n is the length of input sequence. The top layer of architecture evaluates similarity between two final output vector by the two column network. Here, we use cosine similarity as metric.

During the Relation Identification process, the input of left column network is word sequence of question pattern and the input of right column network is sub-relation sequence of certain relation. When conducting question triple matching, we feed complete question word sequence and word sequence of triple into each column network respectively. Note, except for word embedding matrix used in question triple matching is shared between two column networks, other parameters in architecture are independent.

4   Train and Inference

4.1   Training

We adopt margin-based ranking loss function to estimate parameters.

Relation Identification: The training data is denoted by $\mathcal{D} = \{(p_i, r_i) : i = 1, \ldots, |D|\}$, where p_i is the `question pattern` of the ith training question, and the r_i is the corresponding sub-relation label sequences in the knowledge base. The function $S(p, r)$ represents the cosine similarity between the embedded vector of question pattern p and the embedded vector of the relation r, which are the output vectors of the two GRU-based RNN modules (the left and the right columns of our proposed model shown in Fig. 1). This objective function is formulated as:

$$\sum_{i=1}^{|\mathcal{D}|} \sum_{\tilde{r} \in \tilde{R}(r_i)} max\{0, m - S(p_i, r_i) + S(p_i, \tilde{r})\} \qquad (6)$$

where the $\tilde{r}$ is a negative relation for question, which is different with r_i. We will introduce the details of choosing negative examples in the Experiments section. We exploit Adam [13] algorithm to minimize the objective function and to learn the GRU parameters and input embedding vectors.

Question-Triple Matching: We use the same two-column RNN to train the Question-Triple matching model. For this task, the training data is $\mathcal{D} = \{(q_i, t_i) : i = 1, \ldots, |D|\}$, where q_i is the ith training question, and the t_i is one of its correct answer triples in the knowledge base. The word sequences q_i and t_i are the inputs of the left and right columns of our proposed model.

4.2   Inference

Once our model is trained, we can use this model to answer new questions. Given a new question q, by using the entity linking technique proposed in [3], we select the entities whose score is higher than a pre-given threshold as the topic entity set $\mathcal{S}$. Then, we find subgraph of entity $s \in \mathcal{S}$, and extract all one-hop paths or two-hop paths passing CVT node, the relations in the paths are chosen as possible candidate relations. Here, we ignore the first relation in two-hop path. Next, we make use of our learned **Relation Identification** module to discover the top-k (p, r) pair, where k is hyper-parameter and we set it to be 3 in our work. Finally, we find all triples in knowledge base that satisfy the form of $(s, r, ?)$ as candidate triples. We denote this candidate triples as C_q and adopt our **Question-Triple Matching** module to rank them. Because there exists some multi-answer questions, we generate predicated triples set $\widehat{C}_q$ as:

$$\widehat{C}_q = \{\hat{t} | \hat{t} \in C_q \ and \ S(q, \hat{t}) > S(q, t^*) - m\} \tag{7}$$

where $S(q, t^*)$ is the highest score and we use the same threshold m as in Eq. 6.

5 Experiments

We conduct experiments on the WEBQUESTIONS testing set to evaluate our system.

DATASET: WEBQUESTIONS [4] is a popular dataset to evaluate efficiency of QA system, which consists of 5810 question-answer pairs. Because WEBQUESTIONS provides only question-answer pair, we simulate question answering process to collect relation information for training Relation Identification model. Firstly, we use Entity Matching Module described in Sect. 3.2 to get candidate topic entities for questions. Then the 1-hop or 2-hop passing CVT paths on the FREEBASE that connect a candidate topic entity to at least one answer entity are identified as candidate relations. Finally, the relations connecting the most answer entities are voted as correct relations. Other relations founded in QA process are regarded as negative relations.

FREEBASE: As the WEBQUESTIONS dataset uses entities in FREEBASE, we adopt this knowledge base to develop our model. FREEBASE is a large collaborative knowledge base consisting of data composed mainly by its community members. To make FREEBASE fit in memory, we apply the similar preprocess method presented in [7] to extract a subset of FREEBASE.

5.1 Setting

In our experiments, all hyper-parameters are chosen on the WEBQUESTIONS validation set. The size of word vectors d_w, sub-relation vectors d_r and hidden state of GRUs d_g are selected among {64, 128, 192, 256}. We used mini-batch Adam algorithm [13], where batch size is 40, initial learning rate α is selected among {0.1, 0.01, 0.001, 0.0001}. Initial weighs of GRUs are drawn from a 0-mean truncated normal distribution with 0.1 standard deviations. Embedding of word and sub-relation are initialized in same way. The bias inside GRUs are started as 1.0 to make cell not reset and not update. The margin m in Eq. 6 is set to 0.1. Optimal configurations are: $d_w = 192$, $d_r = 192$, $d_g = 192$, $\alpha = 0.001$.

Table 1. Evaluation result on the testing set of WEBQUESTIONS, compared to baselines. The results of baselines are from their original papers.

Method	F1
Berant et al. [4]	31.4%
Berant and Liang [5]	39.9%
Bao et al. [2]	37.5%
Yao and Van Durme [18]	33.0%
Bordes et al. [7]	39.2%
Bordes et al. [9]	29.7%
Dong et al. [11]	40.8%
Our method	**42.0%**

Table 2. Evaluation results of different settings. The listed p@k results of Relation Identification are evaluated on correct relations.

	All	Relation identification
F1	42.0%	40.9%
p@1	43%	61%
p@3	52%	69%
p@5	57%	73.2%

5.2 Experimental Result

We compare our system in terms of average F1 score as computed by the official evaluation script provided by (Berant et al. 2013). For each testing question, we compare the predicted answer set to gold answer set, and compute its F1 score. After going through the whole testing set, we get the popular macro F1 metric that is the average value of the F1 score of all testing samples. As shown in Table 1, our system achieves comparable or better result than baseline system on WEBQUESTIONS.

We also conduct experiments to examine the effect of the core Relation Identification module. Given a question, We applied Relation Identification module to rank its candidate topic entity and relation pair (s, p), As shown in Table 2, the correct (s, p) of 60% questions are ranked at first place. Note that, when only using Relation Identification Module to achieve QA task, all retrieved entities have the same score. Therefore, the listed p@k results of Relation Identification are evaluated on correct relations. After further analyzing, we discover that 170 questions have no corresponding paths between topic entity and answer entities in FREEBASE. Ignoring these 170 questions, the Relation Identification model achieves P@1 = 66.5%, with the average ranking of the correct (s, p) being 4.28, and the average number of candidate (s, p) pairs is 70.7. For a given question, if we take directly the entities connecting to s by p as the predicted answer set, where the (s, q) pair is the first ranked result of the Relation Identification module, we get F1 = 40.9%, which is an acceptable result. This proves that the Relation Identification module achieve a good efficiency.

5.3 Relation Word Detection

In this section, we show how the trained GRU-RNN extracts relation key word from input question pattern. As we know, the GRU layer generates

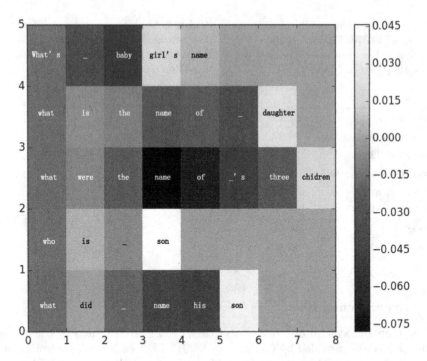

Fig. 2. The 57th dimension of word vectors output by GRU-RNN layer when given 5 question patterns expressing semantic of people.person.children. The vertical axis is the question pattern index from 1 to 5, and the horizontal axis is the word index from 1 to 8 numbered from left to right in a sequence of words, and color codes show activation values. Relation key words like girls, daughter, children, son have relatively high value in each question pattern. (Color figure online)

vector h_t for each input word, the following max-pooling layer takes the maximum value of each vector dimension to form final question pattern representation p. Intuitively, for each dimension of vector, the word with the maximum value contributes the most. We feed some question patterns into GRU-RNN and inspect output vector for each word. A interesting phenomenon is observed that different dimensions of vector are sensitive to different relation words. For instance, the 57th dimension will turn on when meeting words indicating relation people.person.children. And the 43rd dimension will be activated by words indicating relation *people.marriage.spouse*. Figure 2 shows the 57th dimension of word vectors output by GRU-RNN when given 5 following question patterns:

* What's _ baby girl's name?
* What is the name of _ daughter?
* What were the name of _'s three children?
* Who is _ son?
* What did _ name his son?

which all express the same semantic of *people.person.children*. Obviously, words with the maximum value in each question pattern respectively are girls, daughter, children, son, son, which are exactly relation key words.

5.4 Error Analysis

We randomly select some questions from the wrongly answered questions to find out possible causes.

Entity Linking: In entity linking stage, some entity mentions failed to be linked due to POS error. Meanwhile some other entity mentions are correctly located but its corresponding topic entity are dropped due to the low matching score.

Relation Predication: For a part of questions, there do not exist any 1-hop or 2-hop passing CVT node path from its topic entity to answers. As a result, our method can not answer this type of questions for now. What's more, some questions are roughly answered because there is no single relation exactly expressing the semantic of question. For example, we answer *who is Keyshia cole dad* with Keyshia cole's dad and mom based on relation *people.person.parent*, because there is no relation like *people.person.dad*. Overall, most of errors come from incorrect rank of relations.

Constraints and Aggregations: Some questions contain constraint words. For instance, the question *who did jackie robinson first play for*, asking the role that Jackie Robinson played as his first time. Only identifying the relation *sports.sports_team_roster.team* is not sufficient to correctly answer it. Such that, we need further aggregation operation or develop more advanced mechanisms.

Label Error: Some errors in fact are caused by label issues and are not real mistakes. For instance, standard answer set to What are the songs that Justin Bieber wrote only contains 10 songs, which is not completely labeled. And sometimes, the answers that we provide is accepted. For example, we answer *Where did francisco coronado come from* with the entity *Salamance* which is a city northwestern Spain, while the gold answer is *Spain*. What's more, the standard answers of some questions are wrong.

6 Conclusion

In this paper, we propose our knowledge graph based question answering system. We divide the question answering problem into three different parts, and provide three corresponding sub-systems. The GRU-based RNN is the core tool to Relation Identification and Candidates Ranking, because we take the natural language and KB triples as sequences data. Our system achieve comparable result than other baselines, including both semantic parsing and features extraction methods, with a intuitive and simple system structure rather than the complex human handcraft feature or delicate neural network they used.

References

1. Auer, S., Bizer, C., Kobilarov, G., Lehmann, J., Cyganiak, R., Ives, Z.: DBpedia: a nucleus for a web of open data. In: Aberer, K., et al. (eds.) ASWC/ISWC - 2007. LNCS, vol. 4825, pp. 722–735. Springer, Heidelberg (2007). doi:10.1007/978-3-540-76298-0_52
2. Bao, J., Duan, N., Zhou, M., Zhao, T.: Knowledge-based question answering as machine translation. Cell **2**(6) (2014)
3. Bast, H., Haussmann, E.: More accurate question answering on freebase. In: Proceedings of 24th ACM International on Conference on Information and Knowledge Management, pp. 1431–1440. ACM (2015)
4. Berant, J., Chou, A., Frostig, R., Liang, P.: Semantic parsing on freebase from question-answer pairs. In: EMNLP, p. 6 (2013)
5. Berant, J., Liang, P.: Semantic parsing via paraphrasing. In: ACL, no. 1, pp. 1415–1425 (2014)
6. Bollacker, K., Evans, C., Paritosh, P., Sturge, T., Taylor, J.: Freebase: a collaboratively created graph database for structuring human knowledge. In: Proceedings of 2008 ACM SIGMOD International Conference on Management of Data, pp. 1247–1250. ACM (2008)
7. Bordes, A., Chopra, S., Weston, J.: Question answering with subgraph embeddings (2014). arXiv preprint arXiv:1406.3676
8. Bordes, A., Usunier, N., Chopra, S., Weston, J.: Large-scale simple question answering with memory networks (2015). arXiv preprint arXiv:1506.02075
9. Bordes, A., Weston, J., Usunier, N.: Open question answering with weakly supervised embedding models. In: Calders, T., Esposito, F., Hüllermeier, E., Meo, R. (eds.) ECML PKDD 2014. LNCS (LNAI), vol. 8724, pp. 165–180. Springer, Heidelberg (2014). doi:10.1007/978-3-662-44848-9_11
10. Cho, K., Van Merriënboer, B., Gulcehre, C., Bahdanau, D., Bougares, F., Schwenk, H., Bengio, Y.: Learning phrase representations using RNN encoder-decoder for statistical machine translation (2014). arXiv preprint arXiv:1406.1078
11. Dong, L., Wei, F., Zhou, M., Xu, K.: Question answering over freebase with multi-column convolutional neural networks. In: Proceedings of 53rd Annual Meeting of the Association for Computational Linguistics and 7th International Joint Conference on Natural Language Processing, vol. 1, pp. 260–269 (2015)
12. Golub, D., He, X.: Character-level question answering with attention (2016). arXiv preprint arXiv:1604.00727
13. Kingma, D., Ba, J.: Adam: a method for stochastic optimization (2014). arXiv preprint arXiv:1412.6980
14. Liang, P., Jordan, M.I., Klein, D.: Learning dependency-based compositional semantics. In: Proceedings of 49th Annual Meeting of the Association for Computational Linguistics: Human Language Technologies, vol. 1, pp. 590–599. Association for Computational Linguistics (2011)
15. Ling, X., Singh, S., Weld, D.S.: Design challenges for entity linking. Trans. Assoc. Comput. Linguist. **3**, 315–328 (2015)
16. Manning, C.D., Surdeanu, M., Bauer, J., Finkel, J.R., Bethard, S., McClosky, D.: The Stanford CoreNLP natural language processing toolkit. In: ACL (System Demonstrations), pp. 55–60 (2014)
17. Suchanek, F.M., Kasneci, G., Weikum, G.: Yago: a core of semantic knowledge. In: Proceedings of 16th International Conference on World Wide Web, pp. 697–706. ACM (2007)

18. Yao, X., Van Durme, B.: Information extraction over structured data: question answering with freebase. In: ACL, no. 1, pp. 956–966. Citeseer (2014)
19. Yih, W.T., Chang, M.W., He, X., Gao, J.: Semantic parsing via staged query graph generation: question answering with knowledge base. In: Association for Computational Linguistics (ACL) (2015)

Towards Personal Relation Extraction Based on Sentence Pattern Tree

Zhao Jiapeng, Yan Yang, Liu Tingwen[✉], and Shi Jinqiao

Institute of Information Engineering, Chinese Academy of Sciences, Beijing, China
{zhaojiapeng,yanyang9021,liutingwen,shijinqiao}@iie.ac.cn

Abstract. Extracting personal relation triple (S, P, O) from large number of unstructured text is crucial to the construction of knowledge graph, knowledge representation and reasoning of personal relation. Aiming at low accuracy in extracting triples from unstructured text, we present a supervised approach to judge whether extracted triples are correct. The approach need to build a knowledge base which contain peoples attributes first, then a sentence pattern tree is learnt according the people attribute knowledge base and the training data. When training, triples are extracted from the text automatically and labelled whether correct or not manually. Then patterns are constructed layer-by-layer according the position of "triple", "pronoun" and "word" in sentence. At the same time, the correct and error number of triples are recorded on each pattern. When testing, the correctness of triples can be judged by the number recorded in matched patterns. According the test result, our approach does better in the training time, the testing time and the F1-value (76.6%) than the ordinary approach based on feature engineering (75.7%). At last, we make the judgement result of sentence pattern tree as a feature to improve the feature engineering approach (77.5%). In addition, this approach has a better expansibility than the traditional one and has guiding significance to the construction of the training set.

Keywords: Human knowledge graph · Personal relation extraction · Pattern match · Feature extraction

1 Introduction

The foundation work to study peoples network behavior is the human knowledge graph construction. It is crucial to the analysis of related text on the web. The triple(S, P, O) is an important part of the knowledge graph construction. However not only the number of triples extracted by information extraction is huge, but also the precision of extracted triples is difficult to satisfactory. To solve this problem, this paper presents an approach to judge whether extracted triples are correct.

Entity relation is the semantic relation between entities. Automatic Content Extraction (ACE) conference defines the relation extraction as: according the predefine relation type, judging whether the specific semantic relation exists

© Springer Nature Singapore Pte Ltd. 2016
H. Chen et al. (Eds.): CCKS 2016, CCIS 650, pp. 92–103, 2016.
DOI: 10.1007/978-981-10-3168-7_9

or the given relation type is correct. Relation Extraction is one of the most important approaches to get personal relation triples.

We put forward a supervised approach to judge whether the triple is correct. The approach need to build a knowledge base which contain people's attributes first, then a sentence pattern tree is learnt according the people attribute knowledge base and the training data. When training, triples are extracted from the text automatically and labelled whether correct or not manually. Then construct patterns according the position of"triple", "pronoun" and "word" in the sentence by level. At the same time, correct and error number of triples match to patterns are recorded. When testing, the correctness of triples is judged by the number recorded in matched patterns. There is no need for our approach to analysis semantics such as dependency relation and syntax. It could learn a set of patterns automatically according the given training set. When the field changes, it could self-study only by the given training set of the corresponding field. Thanks to the tree structure of our patterns, the efficiency of training and testing are relatively high. When the judging result is wrong, our approach could find error instances timely. It's convenient to analyze the causes of errors. Aiming at the shortcomings of our patterns in not considering person attributes the distance between peoples and relation indicator words and the distinguishing ability of relation indicator words, we extract features of people to improve our pattern approach.

2   Related Work

Prior work on entity relation extraction can be classified to three categories, namely Pattern Match, Semantic Analysis, Feature Classification.

The Pattern Match approach [6,10,11] first formulates the corresponding patterns and relation types according the observation and analysis of instances in training set. Then, match instances in the testing set with patterns preformulated. If any match, we can judge the relation type by the pattern. The main problem of Pattern Match approach is most of patterns are formulated artificially, which make it consume a large amount of human resources. In big data era, the huge scale data makes it impossible to formulate comprehensive and accurate patterns. In addition, when the specific area changes, the original pattern may be wont work well any more. Usually we need to reformulate new patterns to make it adapt to the new area. For example, Qin et al. formulated rules like the relation indicating words must contain verbs to realize the personal relation judging [2]; Paper [5] aiming at the problem of the irrelevant items extraction and missing key information existing in previous work. Through the statistical analysis of the error data, they put forward the approach of making use of part of speech tagging to develop syntactic and lexical constraint patterns to solve the problem; Paper [3] use a semi-supervised approach to extraction. The approach requires manual participation 10–15 min every day. However the target of manual intervention is blind. Not targeted! This paper proposes a judging approach based on Pattern Match, it enumerates various possible situations according the distribution of the training data.

The Semantic Analysis approach deduces some formalized representation which could reflect the meaning of sentence, according to the syntactic structure and the meaning of each notional word in the sentence [13]. By the formalized representation, personal relation could be judged. Using the dependency relation extraction approach, only the part of speech of words is considered, such as paper [7] uses part of speech to formulate patterns. These approaches have no consideration of the semantic gap and usage gap between verbs. For example, both sentence A and sentence B can match some part of speech pattern, but the specific words may lead different meaning; Paper [15] constructs a feature set by the Semantic Role Labeling. Then, a statistical feature combination approach is proposed, and the SVM (Support Vector Machine) classifier is used to realize the semantic analysis; Paper [12] proposes a semantic analysis of noun verbs semantic role labeling based on the traditional verb semantic role labeling. The approach could be used to realize the information extraction; Paper [4] mainly uses the Semantic Role Labeling in the Open Information Extraction. The pattern match approach proposed in this paper doesnt analysis the dependence relation of the sentence, this makes the approach avoid those problem exist in semantic analysis.

The Feature Classification approach judges whether the given personal relation is correct by N-Gram features, word-frequency features [16], TF-IDF features [16], sometimes may also contains some pattern features, semantic analysis features [14, 17] in sentences. Classifier such as SVM [8], maximum entropy, decision tree is taken to transform the judging problem to a binary classification problem. Some approaches [1] also utilize the external resources to improve the accuracy of the relation judging. The problem of the feature engineering approach is: Firstly, the feature space for representing text is in very high dimension. It results in low efficiency of training and testing. Secondly, when the classifying quality is not so good, its hard to discover the concrete instance which is wrong, the only thing we can do is to adjust parameters of the classifier or select new features. Thirdly, when the difference of feature distribution in the training set and testing set is great, the classifying quality is bad. Its hard to build a comparatively complete training data set.

3   Our Approach

In this paper, we mainly focus on 19 types of personal relations, namely campus beauty, rivals in love, teachers, clothing clashing, ex-girlfriend, idol, ambiguous, gossip girl, Hearsay discord, ex-wife, confidante, classmate, wife, separate, carbon copy, friends, agent, fellow-villager, cohabitation. These relations belong to entertainment domain. The reason we choose this domain is that the domain has a rich type of relations.

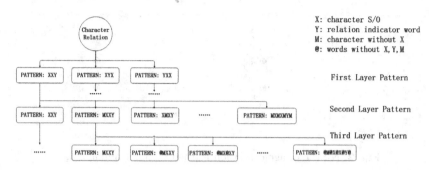

Fig. 1. Overview of N-SPT.

3.1 Selection of Relation Indicating Words

For each kind of relation, we need to find relation indicating words to distinguish them. The number of relation indicating words need to as small as possible and they can represent the 19 kinds of relations effectively.

For a given type, the training data is denoted as $P = \{p_1, p_2, \cdots, p_n\}$, where p_i is text i in the corpus. After segmenting each p_i in P, we can get a dictionary $W = \{w_1, w_2, \cdots, w_m\}$, wi is ith word in the dictionary. Then the selection of relation indicating word could translate into finding the subset $S(S \subseteq W)$ in the dictionary. S should cover P (For each word in p_i, at least one appear in S); S is the minimal set which meet the above conditions, represented as $|S| = \min\{|S_i|\}$, S_i is the subset of all satisfying dictionary. $|*|$ indicates the number of set $*$. Finally, the solved minimal cover of the training set is the relation indicating words.

In a variety of real corpus, there are some high frequency but meaningless words. It makes some meaningful words left in the basket which leads the weight of some keywords reduces. This has a bad influence in the post-processing of the personal relation judgment. For that, we made some adjustments manually.

3.2 N Layers Sentence Pattern Tree (N-SPT)

Construction of N-SPT. For judging the specific relation between peoples by certain sentence, the sentence need contain SPO triple that represent personal relation. Our approach takes the SPO triple consist of peoples and relation indicating words as the core, and increases the number of persons layer-by-layer to extend patterns, which can obtain patterns with hierarchical structure to describe sentences in corpus.

The paper present a kind of N layers Sentence Pattern Tree (N-SPT) based on relation indicating words and personal position relations in sentence and syntactic features, shown in Fig. 1.

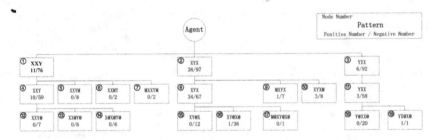

Fig. 2. N-SPT of agent relation learned by our training set.

The first layer of N-SPT only consider the location relation between characters X and relation indicating words Y, which consists of three classes: YXX, XYX, XXY. The location of relation indicating words is crucial to relation judgment.

The second layer of N-SPT considers the influence of third person or personal pronoun M for relation judgment. For each pattern in the first layer, 24 patterns can be generated. For example: for YXX, can generate YXX (not contain the third people), MYXX, YXMX, YXXM, MYMXXM and so on.

The third layer of N-SPT considers @ have an effect on the second layer (Word string @ only considers if there is any word exist, but not consider the specific content and number of words). For example: for MYXX, MYXY (not contain redundant string), M@YXX, MY@XX@, @MY@XX.

For given 19 relations, the paper build a Pattern Tree for each relation. Using the sentences processed in the training set, statically learning Pattern Tree. Parts of the agent relation N-SPT learned by the training set is shown in Fig. 2.

Personal Relation Judgment Based on N-SPT. According the strategy formulated by N-SPT, each sentence will match 3 patterns at most and 1 pattern at least. Using given sentence matches in N-SPT, the positive $PosNumT_i$ and negative number $NegNumT_i$ recorded in the node can be used to Judge the personal relation, as the Eqs. 1 and 2 defined.

$$TP_i = \frac{\min(PosNumT_i, NegNumT_i)}{\max(PosNumT_i, NegNumT_i)} \tag{1}$$

$$TemplateId = \arg\min_i TP_i \quad (1 \le i \le 3) \tag{2}$$

4   Personal Relation Judgment Based on Feature Engineering

In this paper, we set the judging result of N-SPT as one-dimensional feature. Through the analysis of the corpus, we extract some features from text, and use a classifier to judge whether the triple extracted from sentence is correct.

This classifier only consider the position of persons and relation word into consideration, instead of the property of person, the division of relation word, and the distance between persons and relation words. We improved it with classifier of hybrid features. To such sentences which are filtered with the rule of heuristic approach, we extract features from the People Attribute Knowledge Base(PAKB), relation indicating words feature, word-spacing feature as the candidate of the feature classifier approach.

4.1   Feature Extraction Based on People Attribute Knowledge Base(PAKB)

Person Attribute Feature. Aiming at each person in PAKB, including the name, gender, race, height, weight, occupation, the place of birth, registered residence, the date of birth and death, alias and so on, we select all of the above attributes except the name as features. At the same time, we select the number of attributes (not all attributes of a person we can get), the occurrence time of person's name in the training data, the occurrence time of the first and second word of person's name as the candidate feature. In total, we have fifteen features.

Combination Features of Person Attributes. According to person's attribute, the combination of two person properties which need to be determined facilitates the determination of part relation. For instance, if the place of birth or the registered residence is same, the "fellow-townsman" relation is right; whether the gender of two persons is same, the "spouse" relation is wrong. Therefore, we defined four feature combinations as follows:

– Whether the place of birth or registered residence of two persons is same;
– The difference of two persons' age;
– Whether the gender of two persons is same;
– The length of the same prefix of two persons' name.

Feature of Relation Indicating Words. The relation indicating words is got through the approach introduced in Sect. 3.1, and the kind of relation indicating words not only has a low dimension, but also can distinguish 19 types of relations effectively, 72 features in total.

Distance Feature Between Words. For some relations such as "ambiguous", "confidante" and so on, after the analysis of training data, the distance of character and relation word determined the relation is right or not to a certain extent. At the same time, the N-SPT approach doesn't consider the distance feature of persons and relation indicating words. So, we calculate the distance as the candidate feature, SP distance and PO distance, 2 features in total.

4.2   Pattern Tree Features

N-SPT Feature. According to the given sentence, target people and the relation need to be judged. Firstly, we preprocess the sentences and identity whether the target people and relation indicating words are in the sentences or not. If they do not exist, we can judge the relation is error. If exist, we match the sentence with hierarchy. If the sentence matches the pattern, we record the right and wrong numbers and go deep in the next hierarchy; On the contrary, if the sentence could not match, we record the right and wrong number as -1. We set the right and wrong number of patterns as candidate features. In total, we get 6 features in all of the three hierarchies.

N-SPT Result Feature. The effect of N-SPT is very good in the training data. The purpose of using feature classifier approach is to improve the judging effect of N-SPT. In hence, we set the judging result of N-SPT as one of the candidate features.

4.3   Feature Selection

For the selected candidate features, we use entropy formula(formula (4)) to select the best feature for 19 relations, the Entropy(S) is the entropy of collection S, Gain(S,A) is the information gain of sentence collection S, Sv is the collection of correct or error relations.

$$Entropy(S) = -p_+ \log_2 p_+ - p_- \log_2 p_- \tag{3}$$

$$Gain(S, A) = Entropy(S) - \sum_{v \in V(A)} \frac{|S_v|}{|S|} Entropy(S_v) \tag{4}$$

We first choose features with the information gain for each type of relations. Finally, we use the decision tree classifier to judge the personal relations.

5   Experimental Resutls

5.1   Experimental Setup

In our experiments, we denote subject entity as S, object entity as O, and predicate as P. Note that here predicate is the type of personal relations. We evaluate our work on the corpus published by Baidu. There are 7813 labelled samples in the training set, and 2610 unlabelled samples in the testing set. Each sample consists of a sentence, a SPO where entity S and entity O appear in the sentence and predicate P is one of 19 personal relation types mentioned in Sect. 2. A human knowledge base of 12150 persons is also published at the same time, which contains person attributes for each entity in the SPO. In total there are 13 different person attributes in the published human knowledge base, such as person name, sex, age.

We use precision, recall, F1 value as the evaluation index. The formulas are shown as follows:

$$precision_i = \frac{|\bigcap(predictionspo_i, referencespo_i)|}{|predictionspo_i|} \quad (5)$$

$$recall_i = \frac{|\bigcap(predictionspo_i, referencespo_i)|}{|referencespo_i|} \quad (6)$$

$$F1_i = \frac{2 \times precision_i \times recall_i}{precision_i + recall_i} \quad (7)$$

$$F1 = \frac{\sum_{i=1}^{n} F1_i}{n} \quad (8)$$

Meanwhile, $predictionspo_i$ is the number of SPO sets belong to ith relations judged by our approach, $referencespo_i$ is the number of SPO sets really belong to i-th relations, n represents the 19 kinds of relations. We use F1 as the evaluation standard.

5.2 Experiments and Results

Firstly, we preprocessed the sentences and removed the stop words and signals, and keep some important signals such as "book title" mark and double quotation marks. We use person's ID to search the name, because the name is not unique. According to the analysis of sentences, we made some heuristic rules to assist the judgments of relations.

- If the sentence does not contain the relation indicating words, the relation is error.
- If the given name with adjacent word is another name, the relation is error.
- If the given name or relation is contained by signals, the relation is error.
- If the given name exists and its friends and relatives exist, the relation is error.

For some filtered sentences, we firstly get the relation indicating word. Then judging persons relations by N-SPT and use the 7813 sentences as training data, 2610 sentences as test data. We can achieve a F1 value 76.63%.

In Sect. 2.3, we get the candidate features, and in 7813 items of the training data, we use cross-validation approach to select the best feature (the information entropy more than 0.01). We use the decision tree provided by WEKA [9] to judge the personal relation (WDec classifier), F1 value is about 77.506%. Table 1 compared the experiment result of N-SPT and WDec classifier in detail.

We use N-SPT to judge the personal relation directly. The F1 value is about 76.63%. WDec is 77.506%. We also compared the time of training, testing and F1-value (as shown in Table 2). The result demonstrates that our approach is better than the BestResult [18] in this data set.

Table 1. Result of the 19 relations Judge by N-SPT and WDec Classifier

Relation type	N-SPT/WDec of precision (%)	N-SPT/WDec of recall (%)	N-SPT/WDec F1-value (%)
Campus beauty	77.1/77.1	96.4/96.4	85.7/85.7
Rivals in love	85.7/85.7	85.7/85.7	85.7/85.7
Teachers	67.4/71.0	54.7/51.9	60.4/60.0
Clothing clashing	81.1/81.1	88.2/88.2	84.5/84.5
Ex-girlfriend	65.2/65.2	100/100	78.9/78.9
Idol	76.3/76.3	85.5/86.6	80.6/81.1
Ambiguous	78.4/78.4	80.0/81.6	79.2/80.0
Gossip girl	88.0/88.0	84.6/84.6	86.3/86.3
Hearsay discord	73.5/59.4	67.6/67.6	70.4/68.5
Ex-wife	87.5/87.5	77.8/77.8	82.4/82.4
Confidante	69.2/69.2	36.0/37.5	47.4/48.6
Wife	73.2/73.2	87.2/91.1	79.6/81.2
Friends	57.4/71.1	79.5/69.2	66.7/70.1
Separate	74.1/76.9	66.7/66.7	70.2/71.4
Carbon copy	71.4/71.4	71.4/71.4	71.4/71.4
Classmate	87.5/87.5	63.6/63.6	73.7/73.7
Agent	80.0/80.0	85.7/92.3	82.8/85.7
Fellow-villager	69.7/79.3	88.5/92.0	78.0/85.2
Cohabitation	90.6/90.6	93.5/93.5	92.1/92.1
Total			76.6/77.5

Table 2. Comparison of time cost in and F1 value between different classifiers

Approach	Training time	Testing time	F1-Value
WDec Classifier	6.1 min	1.4 min	77.51%
N-SPT	1.4 min	0.45 min	76.63%
BestResult	about 30 min	about 30 min	75.68%

5.3 Experiment Analysis

The experiment use N-SPT to judge whether the personal relation is correct. The training time, testing time and F1-value are all better than BestResult in the dataset. It demonstrates N-SPT can judge the personal relation efficiently and accurately. The reason that our approach achieves desired results and disadvantages of our method are shown as follows:

– BestResult uses N-Gram features, dependency tree features to judge personal relation. It has a very high dimension which leads the training time and testing time consume too much. But in our approach, the N-SPT proposed by us has a good summary of the whole corpus and the feature dimension is rather low. This improves the efficiency of training and testing.

- According to the experiment result, WDec has advantage in 17 kinds of relations which proved that the PAKB feature, relation indicating words feature, the distance of words make up the disadvantage of N-SPT. The improve strategy is effective.
- N-SPT has guiding significance to the construction of training set, as shown in Fig. 2. When the N-SPT is not complete, for example, in node. The layer is less than 3, and we can add some sentences which match the sub-node of to the training data. The quality of training data will be improved by completing the N-SPT.
- Comparing with the word-bag model, N-SPT can locate the wrong instance efficiently and adjust accordingly.

6   Conclusions and Future Work

6.1   Conclusions

Our approach is a supervised approach using the training data to construct the tree patterns. Comparing with traditional works, our approach can construct patterns of the entire training data with little manual participation. When the domain changes, we just need to adjust some coefficient to construct new patterns. In the retrieval and restoration, the efficiency of training and testing are high relatively. However, N-SPT just consider the tree structure in generating the pattern and ignore the attributes of people. Pointed at this fault, we added PAKB, combination features of persons' attributes, relation indicating word and so on. The experiment result shows these improvements have good effects.

6.2   Future Work

The N-SPT presented in this paper works well while processing sentences with concise and simple structures, but it still needs improvement when handling more complex sentences, It still leads to relatively big error when matched to the third level of N-SPT, but the current N-SPT only has three levels, so it can be treated with clustering approaches, such as K-MEANS, hierarchical clustering and LDA, clustering the words in the rest person strings @ on the third level template, and clustering the words which affect relation determination into particular category, thus expending it into the fourth or even deeper levels. N-SPT is highly extensible, so the next focus of this paper will be how to extend N-SPT to even deeper levels in order to process complex sentences. The current way of constructing N-SPT with training is to build a N-SPT for each relation, but there might be several reference words for each relation, and the different usage of each reference word might result in the error of building N-SPT for relations, While building a N-SPT for each reference word might also cause data sparsity problem, so further research is required in order to balance the difference usage of the reference words for relations of N-SPT and the data sparsity problem. This paper only tried decision tree to categorize different combined features at present, different classifiers will be used to test their effect on relation determination in follow-up researches.

Acknowledgement. Supported in part by the Strategic Priority Research Program of the Chinese Academy of Sciences under Grant No. XDA06030200.

References

1. Apostolova, E., Tomuro, N.: Combining visual and textual features for information extraction from online flyers. In: EMNLP, pp. 1924–1929 (2014)
2. Bing, Q., Ting, L.A.L.: Unsupervised Chinese open entity relation extraction. J. Comput. Res. Dev. **52**(5), 1029 (2015)
3. Carlson, A., Betteridge, J., Kisiel, B., Settles, B., Hruschka Jr., E.R., Mitchell, T.M.: Toward an architecture for never-ending language learning. In: AAAI, vol. 5, p. 3 (2010)
4. Christensen, J., Soderland, S., Etzioni, O., et al.: An analysis of open information extraction based on semantic role labeling. In: Proceedings of the Sixth International Conference On Knowledge Capture, pp. 113–120 (2011)
5. Etzioni, O., Fader, A., Christensen, J., Soderland, S., Mausam, M.: Open information extraction: the second generation. In: IJCAI, vol. 11, pp. 3-10 (2011)
6. Fang, Y., Chang, K.C.C.: Searching patterns for relation extraction over the web: rediscovering the pattern-relation duality. In: Proceedings of the Fourth ACM International Conference on Web Search and Data Mining, pp. 825–834 (2011)
7. Gamallo, P., Garcia, M., Fernández-Lanza, S.: Dependency-based open information extraction. In: Proceedings of the Joint Workshop on Unsupervised and Semi-supervised Learning in NLP, pp. 10–18 (2012)
8. Glass, M., Barker, K.: Bootstrapping relation extraction using parallel news articles. In: Proceedings of the IJCAI Workshop on Learning by Reading and its Applications in Intelligent Question-answering, Barcelona (2011)
9. Hall, M., Frank, E., Holmes, G., Pfahringer, B., Reutemann, P., Witten, I.H.: The weka data mining software: an update. ACM SIGKDD Explor. Newslett. **11**(1), 10–18 (2009)
10. Kluegl, P., Toepfer, M., Beck, P.D., Fette, G., Puppe, F.: Uima ruta: rapid development of rule-based information extraction applications. Nat. Lang. Eng. **22**(01), 1–40 (2016)
11. Kozareva, Z., Hovy, E.: Learning arguments and supertypes of semantic relations using recursive patterns. In: Proceedings of the 48th Annual Meeting of the Association for Computational Linguistics, pp. 1482–1491 (2010)
12. Li, J.H., Zhou, G.D., Zhu, Q.M., Qian, P.D.: Semantic role labeling in Chinese language for nominal predicates. Ruanjian Xuebao/J. Softw. **22**(8), 1725–1737 (2011)
13. Lim, S., Lee, C., Ra, D.: Dependency-based semantic role labeling using sequence labeling with a structural SVM. Pattern Recogn. Lett. **34**(6), 696–702 (2013)
14. Nie, T., Shen, D., Kou, Y., Yu, G., Yue, D.: An entity relation extraction model based on semantic pattern matching. In: 2011 Eighth Web Information Systems and Applications Conference (WISA), pp. 7–12 (2011)

15. Sq, L., Tj, Z., Hj, L., Py, L.: Chinese semantic role labeling based on feature combination. J. Softw. **22**(2), 222–232 (2011)
16. Sun, A., Grishman, R., Sekine, S.: Semi-supervised relation extraction with large-scale word clustering. In: Proceedings of the 49th Annual Meeting of the Association for Computational Linguistics: Human Language Technologies, vol. 1, pp. 521–529 (2011)
17. Zahedi, M.H., Kahani, M.: SREC: discourse-level semantic relation extraction from text. Neural Comput. Appl. **23**(6), 1573–1582 (2013)
18. Zhihua, Z., Jianxiang, W., Junfeng, T., Guoshun, W., Man, L.: Blocked person relation recognition system based on multiple features. J. Comput. Appl. **36**(3), 751 (2016)

An Initial Ingredient Analysis of Drugs Approved by China Food and Drug Administration

Haodi Li, Qingcai Chen, Buzhou Tang[✉], Dong Huang,
Xiaolong Wang, and Zengjian Liu

Key Laboratory of Network Oriented Intelligent Computation,
Shenzhen Graduate School, Harbin Institute of Technology,
518055 Shenzhen, China
haodili.hit@gmail.com, qingcai.chen@gmail.com,
tangbuzhou@gmail.com, donghuang2012@gmail.com,
wangxl@insun.hit.edu.cn, liuzengjian.hit@gmail.com

Abstract. Drug is an important part of medicine. Drug knowledge bases that organize and manage drugs have attracted considerable attention, and have been widely used in human health care in many countries and regions. There are also a large number of electronic drug knowledge bases publicly available. In China, however, there is hardly any publicly available well-structured drug knowledge base, may due to two different types of medicine: Chinese traditional medicine (CTM) and modern medicine (ME). In order to build an electronic knowledge base of drugs approved by China Food and Drug Administration (CFDA), we developed a preliminary ingredient drug analysis system. This system collects all drug names from the website of CFDA, obtains their manuals from three medical websites, extracts the ingredients of drugs, and analyses the distribution of the extracted ingredients. Totally, 12,918 out of 19,490 drug manuals were collected. Evaluation on randomly selected 50 drug manuals shows that the system achieves an F-score of 95.46% on ingredient extraction. According to the distribution of the extraction ingredients, we find that ingredient multiplexing is very common in medicine, especially in herbal medicine, which may provide a clue for drug safety as taking more than one type of drug that contains partially the same ingredients may cause overtaking the same ingredients.

Keywords: Drug knowledge base · Chinese traditional medicine · Drug ingredient extraction

1 Introduction

In human's history, medicine always attracts considerable attention. Until now, it has made great progress and various types of medicine with different types of drugs appear such as Chinese traditional medicine (CTM) and modern medicine (ME). In a country or region, there may be more than one type of medicine. For example, in China, CTM and ME coexist. Most drugs in ME consist of only one chemical substance, while most drugs in CTM consists of multiple medicinal herbs. The elementary units of drugs in

© Springer Nature Singapore Pte Ltd. 2016
H. Chen et al. (Eds.): CCKS 2016, CCIS 650, pp. 104–109, 2016.
DOI: 10.1007/978-981-10-3168-7_10

ME are different from that of drugs in CTM. For drugs in ME, there have been a large number of public electronic knowledge bases in the United States of America (USA), which have been widely used in human health care. However, few electronic knowledge bases of drugs in other types of medicine such as CTM are available.

In order to build a well-structured electronic knowledge base of drugs approved by China FDA (CFDA), we collect all drug names from the website of CFDA (http://www.sda.gov.cn), obtain their manuals from some medical websites, and analyse them briefly. Among these drugs, manuals of 12,918 drugs are collected from medical websites. In order to analyse the drugs, we build an automatic ingredient extraction system based on manuals of 320 randomly selected out of the 12,918 drugs. Evaluation on manuals of the other randomly selected 50 drugs shows that the ingredient extraction system achieves a precision of 96.51%, a recall of 94.44% and an F-score of 95.46%. With this system, all ingredients are extracted from the 12,918 manuals. Based on the extracted ingredients, we find that the ingredient multiplexing is very common in medicine, especially in herbal medicine.

2 Related Work

A large number of drug knowledge bases have been developed for different applications such as medication information exchange, clinical decision support, etc. In the USA, both government departments and academically institutions have been involved in building and maintaining various types of drug knowledge bases. The representative drug knowledge bases include the FDA Terminology, NDF-RT [1], RxNorm [2], DrugBank [3], medical databases in UMLS [4] and so on. The FDA Terminology is developed by US FDA and used to support medication information exchange between government agencies by using the Unique Ingredient Identifier (UNII) codes that uniquely identify all ingredients of marketed drugs in the USA to control terminology in medication information area. NDF-RT is a drug database made and maintained by the Veterans Health Administration (VHA). RxNorm provides normalized names for clinical drugs and links them to many drug vocabularies and databases. DrugBank is a database of FDA-approved drugs, nutraceuticals and experimental drugs. UMLS is developed and maintained by the US National Library of Medicine (NLM). It indexes and links various dictionaries through a simple semantic network. All these drug knowledge bases are digitized and most of them are publicly available.

In China, the related research on drug knowledge base construction started later. The early studies mainly focused on how to interpret each term of drug dictionaries. For example, the Chinese Pharmacopoeia edited by the National Pharmacopoeia Committee of China uses the active ingredients of drugs as the basic units to describe the drugs' chemical structure, properties, detect methods and so on [5]. The Dictionary of Chinese Pharmacy uses drug ingredients as the basic units to describe drugs' aliases and comments [6]. The Contemporary Drug's Names and Tradenames Dictionary edited by the China Association of Traditional Chinese Medicine also uses the active ingredients of drugs as the basic units to describe the drugs' category, relative diseases, aliases and production name [7]. In recent years, some researchers have begun to use semantic relations to construct drug knowledge bases such as the Traditional Chinese

Medicine Language System developed by the China Association of Traditional Chinese Medicine [8]. Most of drug knowledge bases only focus on drugs in herbal medicine, and there is hardly any publicly available electronic drug knowledge base.

3 Method

Figure 1 shows the overview of our preliminary drug ingredient analysis system. It consists of five components as follows:

(1) Drug Name Extraction: extract drug names from the CDFA website (http://www.sda.gov.cn) by a customized crawler. 19,490 drugs have been approved by CFDA in total until January 2015, and have been classified into seven categories: herbal medicine (9,914), chemical medicine (8,879), accessory (30), biologicals (555), pharmaceutic adjuvant (6) and other (106).

(2) Drug Manual Collection: collect drug manuals from three medical websites, i.e., http://ypk.39.net, http://www.yaopinnet.com and http://db.yaozh.com. We collect all manuals in text form, and finally obtain 16,882 manuals.

(3) Ingredient Annotation: randomly select 370 drug manuals for annotation. Among them, 320 manuals are used as a training set, and the reminding 50 manuals are used as a test set.

(4) Ingredient Extraction: extract ingredients of drugs from their manuals. This task is recognized as a sequence labeling problem, and Conditional Random Fields (CRF) is used to solve it. The first step of ingredient extraction is to split every manual into sentences. After sentence split and tokenization, each ingredient is represented by BILO tags, where B, I, L and O denote a Chinese character at the beginning, in the middle, at the ending and outside of an ingredient respectively. An example of the ingredient representation is shown in Fig. 2. A CRF model is trained on the training set, and all collected drug manuals are labeled by the model. The features used in the CRF-based system only include N-grams of tokens ($N = 1, 2, 3$ in a window of $[-3, 3]$), segmentation and part-of-speech.

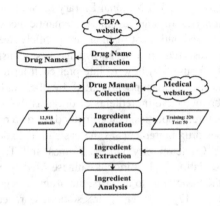

Fig. 1. Overview of our preliminary drug ingredient analysis system.

Drug Name: "布拉氏酵母菌散" (Saccharomyces boulardii sachets)

Ingredient Statement:

"本品主要活性成份：冻干布拉氏酵母菌。辅料：果糖、乳糖、微粉硅胶、水果味香精。"

BILO Representation:

"本/O 品/O 主/O 要/O 活/O 性/O 成/O 份/O：/O 冻/B 干/I 布/I 拉/I 氏/I 酵/I 母/I 菌/I。/O 辅/O 料/O：/O 果/O 糖/O、/O 乳/O 糖/O、/O 微/O 粉/O 硅/O 胶/O、/O 水/O 果/O 味/O 香/O 精/O。/O "

Fig. 2. Example of the ingredient representation.

Precision, recall and F-score are used to measure the performance of the ingredient extraction system.

(5) Ingredient Analysis: analyse the distribution of ingredients in the drugs approved by CFDA according to the results of the ingredient extraction module.

4 Result

The precision, recall, F-score of our ingredient extraction system on the test set are 96.51%, 94.44% and 95.46% respectively. On the 25 drug manuals in herbal medicine, the ingredient extraction system achieves a precision of 96.47%, a recall of 95.00% and an F-score of 95.71% respectively, while it achieves a precision of 96.88%, a recall of 91.18% and an F-score of 93.94% on the 25 drug manuals in chemical medicine. Obviously, the ingredient extraction system shows better performance in herbal medicine than chemical medicine.

On all 12,918 drug manuals in text format, the ingredient extraction system obtains 5,107 types of ingredients, including 3,420 types of herbal ingredients and 2,102 types of chemical ingredients. To further understand the distribution of the ingredients, we list the most common 5 ingredients of drugs in herbal medicine in Table 1 and the most 5 ingredients of drugs in chemical medicine in Table 2 respectively. The most common ingredient of drugs in herbal medicine is liquorice ("甘草" in Chinese), which occurs in 1,659 drugs, and the most common ingredient of drugs in chemical medicine is acetaminophen ("对乙酰氨基酚" in Chinese), which occurs in 158 drugs. It seems that ingredient multiplexing in herbal medicine is more common than chemical medicine. To validate it, we further investigate the relationship between the number of drugs and

Table 1. Most common 5 ingredients of drugs in herbal medicine

Ingredients		Counts
Name	Chinese name	
Liquorice	甘草	1716
Angelica sinensis	当归	1556
Astragalus membranaceus	黄芪	1081
Poria cocos	茯苓	1069
Ligusticum chuanxiong hort.	川芎	927

the number of ingredients as shown in Fig. 3, where x axis is the number of ingredients sorted by the count they occur in drugs and y axis is the number of drugs containing the corresponding ingredient. It is clear that the ingredient multiplexing in herbal medicine is more common than chemical medicine.

Table 2. Most common 5 ingredients of drugs in chemical medicine

Ingredients		Counts
Name	Chinese name	
Acetaminophen	对乙酰氨基酚	158
Chlorpheniramine maleate	马来酸氯苯那敏	151
Vitamin B	维生素B	145
Glycerin	甘油	121
Sodium chloride	氯化钠	116

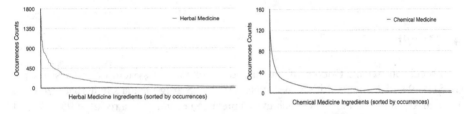

Fig. 3. Relationship between the number of drugs and the number of ingredients.

5   Discussions and Conclusion

In this study, we analyse the distributions of ingredients of drugs approved by CFDA, where the ingredients are extracted by a CRF-based classifier. As the CRF-based ingredient extraction system achieves an F-score of 95.46% on an independent test set, the analysis would be worthy of trust.

We notice that a number of drug manuals (6,572 out of 19,490) cannot be collected from the three medical websites. Most of them are not available on the internet, and a small number of them are only available in non-plain text format. Therefore, drug information needs to be further digitized. In our future work, we will manually add the missed drug manuals to our database.

Although the manuals of drugs are well formatted, it is not easy to extract ingredients from them by simple rules. At the beginning of this study, we have ever attempted to extract ingredients from the first sentences in the "ingredients" field of drug manuals by splitting the sentences by punctuations and treating each part as an ingredient. However, this rule-based method achieves only a precision of 84.68%, a recall of 85.04% and an F-score of 84.86% on the test set, which are much worse than the CRF-based classifier. The main challenge lies in that the ingredients of some drugs are not given directly.

There are some interesting findings from the extracted ingredients. Firstly, two drugs may have the same ingredients such as "JuBanZhiKe Granule" and "JuHong

Pill" ("橘半止咳颗粒" and "橘红丸" in Chinese), both of which consist of 14 herbs. Secondly, one drug may consist of a subset of ingredients of another drug. For example, "ShaYao" ("痧药" in Chinese), a drug in herbal medicine, is composed of 11 herbs, and another drug in herbal medicine "ChanSuDing" ("蟾酥锭" in Chinese) is composed of 4 out of the 11 herbs of "ShaYao". These two drugs look similar according to their ingredients but their indications are greatly different.

This study is a preliminary step of other studies such as medical knowledge graph construction, but it is can be widely used in several medical applications. It may guide the suitable usage of drugs. For example, drugs that contain the same ingredients had better be taken separately as overdosing one ingredient may cause potential side effects such as polygonum multiflorum [9] ("何首乌" in Chinese).

Ingredient multiplexing is very common in medicine, especially in herbal medicine. Based on the results of the drug ingredient extraction system, we can further link drugs through their common ingredient(s), which is a part of knowledge graph of drugs and is one case of our future work.

Acknowledgments. This paper is supported in part by grants: National 863 Program of China (2015AA015405), NSFCs (National Natural Science Foundation of China) (61402128, 61473101, 61173075 and 61272383) and Strategic Emerging Industry Development Special Funds of Shenzhen (JCYJ20140508161040764, JCYJ20140417172417105 and JCYJ20140627163809422).

References

1. Brown, S.H., Elkin, P.L., Rosenbloom, S.T., Husser, C., Bauer, B.A., Lincoln, M.J., Carter, J., Erlbaum, M., Tuttle, M.S.: VA national drug file reference terminology: a cross-institutional content coverage study. Medinfo **11**, 477–481 (2004)
2. Liu, S., Ma, W., Moore, R., Ganesan, V., Nelson, S.: RxNorm: prescription for electronic drug information exchange. IT Prof. **7**, 17–23 (2005)
3. Law, V., Knox, C., Djoumbou, Y., Jewison, T., Guo, A.C., Liu, Y., Maciejewski, A., Arndt, D., Wilson, M., Neveu, V., et al.: DrugBank 4.0: shedding new light on drug metabolism. Nucleic Acids Res. **42**, D1091–D1097 (2014)
4. Bodenreider, O.: The unified medical language system (UMLS): integrating biomedical terminology. Nucleic Acids Res. **32**, D267–D270 (2004)
5. C.P. Commission, et al.: Chinese Pharmacopoeia, vol. 328, p. 547. Chemical Industry Press, Beijing (2005)
6. Xie, G.: Zhōngyī Dàcídiǎn. People's Health Publisher, Beijing (1998)
7. Zhao, Z.: Contemporal Drug's Names and Tradenames Dictionary. Chemical Industry Press, Beijing (2006)
8. Yi, A.-N., Zhang, N.: A study on unified traditional chinese medicine language system. Chin. J. Inf. Tradit. Chin. Med. **10**, 90–92 (2003)
9. Lei, X., Chen, J., Ren, J., Li, Y., Zhai, J., Mu, W., Zhang, L., Zheng, W., Tian, G., Shang, H.: Liver damage associated with polygonum multiflorum thunb.: a systematic review of case reports and case series. Evid.-Based Complement. Altern. Med.: eCAM **2015**, 459–749 (2015)

A Tableau-Based Forgetting in ALCQ

Hong Fang[1] and Xiaowang Zhang[2,3,4(✉)]

[1] College of Arts and Sciences, Shanghai Polytechnic University,
Shanghai 201209, China
[2] School of Computer Science and Technology,
Tianjin University, Tianjin 300350, China
xiaowangzhang@tju.edu.cn
[3] Tianjin Key Laboratory of Cognitive Computing and Application,
Tianjin 300350, China
[4] Key Laboratory of Computer Network and Information Integration,
Southeast University, Ministry of Education, Nanjing 211189, China

Abstract. Forgetting is a useful tool for tailoring ontologies by reducing the number of concepts and roles. The issue of forgetting for general ontologies in more expressive description logics, such as $\mathcal{ALCQ}$ and $\mathcal{SHIQ}$, is largely unexplored. In this paper, we develop a decidable, sound, and complete tableau-based algorithm to implement the forgetting-based reasoning. Our tableau algorithm is technically feasibly extended to explore the forgetting in more expressive ontology languages.

1 Introduction

The Semantic Web [1], as an extension of the World Wide Web (WWW), becomes more constantly changing and highly collaborative. Ontologies in Semantic Web can be used by automated tools to provide advanced services such as more accurate web search, intelligent software agents and knowledge management. An example of large biomedical ontology is SNOMED CT. Ontology editing and maintaining tools, such as Protégé, are supported by efficient reasoners based on tableau algorithms for description logics (DLs) [1]. However, as shown in [1], the existing reasoners provide limited reasoning supports for ontology modifications, which largely restricts the wide use of ontologies in the Semantic Web.

Forgetting [3], as an important tool for tailoring ontologies by reducing the number of concepts and roles [3]. It is proven that forgetting can be applied in ontology revision [3], ontology repair [5], and ontology reasoning [6] etc. Though there are some approaches to characterize the forgetting-based reasoning over ontologies [5], it is still interesting to develop some algorithm to characterize the forgetting-based reasoning.

Moreover, it is also interesting to develop some approaches to computing the results of forgetting over ontologies. Recently, there exist some works addressed this issue. For instance, a rewriting approach is presented to compute uniform interpolation in DL-Lite. However, this approach is not direct to treat ontologies

© Springer Nature Singapore Pte Ltd. 2016
H. Chen et al. (Eds.): CCKS 2016, CCIS 650, pp. 110–116, 2016.
DOI: 10.1007/978-981-10-3168-7_11

in expressive description logics even basic description logic $\mathcal{ALC}$. As an attempt, Wang et al. [3] have firstly defined semantic forgetting about concepts and roles in $\mathcal{ALC}$ ontologies and have presented an algorithm to computing the result of forgetting where all concepts are required in disjunctive norm form (DNF). In [4], a tableau-based approach is proposed to compute the results of forgetting over $\mathcal{ALC}$ ontologies where concepts are required in negation normal form (NNF) instead of DNF.

In this paper, inspired from [4], we extend this tableau-based approach to characterize forgetting-based reasoning and generate the rolling-up technique to compute the result of forgetting over ontologies in expressive description logics. This paper focuses the description logic $\mathcal{ALCQ}$ since the number restriction $\mathcal{Q}$ is a most expressive operator in constructing many expressive description logics $\mathcal{SHIQ}$ [2]. Compared with the tableau-based approach introduced in [4], our proposal can further treat ontologies with the number restriction $\mathcal{Q}$.

2   Preliminaries

In this section, we briefly recall some preliminaries of $\mathcal{ALCQ}$ and the tableau algorithm for reasoning tasks. Further details of $\mathcal{ALCQ}$ and the tableau algorithm for $\mathcal{ALCQ}$ can be found in [1,2].

Description Logic $\mathcal{ALCQ}$. First, we introduce the syntax of *concept descriptions* for $\mathcal{ALCQ}$. To this end, we assume that N_C is a set of *concept names*, N_R is a set of *role names* and N_I is a set of individuals.

Elementary concept descriptions consist of *concept names* and *role names*. So a concept name is also called *atomic concept* while a role name is also called *atomic role*.

Concepts description in $\mathcal{ALCQ}$ can be formed according to the following syntax:

$$C, D \rightarrow A \mid \top \mid \bot \mid \neg C \mid C \sqcap D \mid C \sqcup D \mid \forall R.C \mid \exists R.C \mid \leq nR.C \mid \geq nR.C$$

An interpretation $\mathcal{I}$ of $\mathcal{ALCQ}$ is a pair $(\Delta^{\mathcal{I}}, \cdot^{\mathcal{I}})$ where $\Delta^{\mathcal{I}}$ is a non-empty set called the *domain* and $\cdot^{\mathcal{I}}$ is an interpretation function which associates each atomic concept A with a subset $A^{\mathcal{I}}$ of $\Delta^{\mathcal{I}}$ and each role R with a binary relation $R^{\mathcal{I}} \subseteq \Delta^{\mathcal{I}} \times \Delta^{\mathcal{I}}$. This function $\cdot^{\mathcal{I}}$ can be naturally extended to complex descriptions as normal [1].

An *assertional box* (or *ABox*) is a finite set of *assertions*. An assertion is a *concept assertion* of the form $C(a)$ or a *role assertion* of the form $R(a, b)$, where a and b are individuals, C is a concept and R is a role. An interpretation $\mathcal{I}$ *satisfies* a concept assertion $C(a)$ if $a^{\mathcal{I}} \in C^{\mathcal{I}}$, a role assertion $R(a, b)$ if $(a^{\mathcal{I}}, b^{\mathcal{I}}) \in R^{\mathcal{I}}$. If an assertion ϕ, it is denoted $\mathcal{I} \models \phi$. An interpretation $\mathcal{I}$ is a *model* of an ABox $\mathcal{A}$, denoted by $\mathcal{I} \models \mathcal{A}$, if it satisfied all assertions in $\mathcal{A}$.

An *inclusion axiom* (simply *inclusion*, or *axiom*) is of the form $C \sqsubseteq D$ (C is *subsumed* by D), where C and D are concept descriptions. The inclusion $C \equiv D$ (C is *equivalent* to D) is an abbreviation of two inclusions $C \sqsubseteq D$ and $D \sqsubseteq C$.

A *terminology box*, or *TBox*, is a finite set of inclusions. An interpretation $\mathcal{I}$ satisfies an inclusion $C \sqsubseteq D$ if $C^{\mathcal{I}} \subseteq D^{\mathcal{I}}$. $\mathcal{I}$ is a *model* of a TBox $\mathcal{T}$, denoted by $\mathcal{I} \models \mathcal{T}$, if $\mathcal{I}$ satisfies every inclusion of $\mathcal{T}$.

Formally, an *ontology* $\mathcal{O}$ is a pair $(\mathcal{T}, \mathcal{A})$ of a TBox $\mathcal{T}$ and an ABox $\mathcal{A}$. An interpretation $\mathcal{I}$ is a *model* of $\mathcal{O}$ if $\mathcal{I}$ is a model of both $\mathcal{T}$ and $\mathcal{A}$, denoted by $\mathcal{I} \models \mathcal{O}$. If ϕ is an axiom or an assertion, an ontology $\mathcal{O}$ *entails* ϕ, denoted by $\mathcal{O} \models \phi$, if every model of $\mathcal{O}$ is also a model of ϕ. Two ontologies $\mathcal{O}$ and $\mathcal{O}'$ are *equivalent*, denoted by $\mathcal{O} \equiv \mathcal{O}'$, if they have the same models. The equivalent relationship "$\equiv$" can be similarly defined for ABoxes and TBoxes.

The signature of a concept description C, written $sig(C)$, is the set of all concept names and role names in C. Similarly, we can define $sig(\mathcal{A})$ for an ABox $\mathcal{A}$, $sig(\mathcal{T})$ for a TBox $\mathcal{T}$, and $sig(\mathcal{O})$ for an ontology $\mathcal{O}$.

Tableau-Based Reasoning in $\mathcal{ALCQ}$. The tableau based algorithms have been developed to decide the consistency of general DL ontologies.

Given an ontology $\mathcal{O} = (\mathcal{T}, \mathcal{A})$, we can assume without loss of generality that all of the concepts occurring in $\mathcal{T}$ and $\mathcal{A}$ are in NNF, i.e., that negation ($\neg$) is always in front of concept names. Note that an arbitrary $\mathcal{ALCQ}$ concept can be transformed into an equivalent one in NNF in polynomial time by applying the following rules:

$$\neg(C \sqcup D) \equiv \neg C \sqcap \neg D, \quad \neg \forall R.C \equiv \exists R.\neg C, \quad \neg \geq nR.C \equiv \, \leq n-1R.C,$$
$$\neg(C \sqcap D) \equiv \neg C \sqcup \neg D, \quad \neg \exists R.C \equiv \forall R.\neg C, \quad \neg \leq nR.C \equiv \, \geq n+1R.C.$$

where $\leq (-1)R.C \equiv A \sqcap \neg A$ for some $A \in N_C$. Given a concept C, we use $\dot{\neg} C$ to denote the NNF of $\neg C$.

The tableau algorithm works on a data structure called a *completion forest*. This consists a labeled directed graph, each node of which is the root of a *completion tree*. Each node x is labeled a set of concepts $\mathcal{L}(x)$ and each edge $\langle x, y \rangle$ is labeled a set of roles $\mathcal{L}(\langle x, y \rangle)$. If a role $R \in \mathcal{L}(\langle x, y \rangle)$, then we say x is an *R-predecessor* of y (and that y is an *R-successor* of x). A node y is an *ancestor* of a node x if they both belong to the same completion tree and either y is a predecessor of x, or there exists a predecessor z of x such that y is an ancestor of z.

Firstly, the completion forest is initialized $\mathcal{F}$ such that is contains a root node x_a, with $\mathcal{L}(x_a) = \{C \mid a : C \in \mathcal{A}\}$ for each individual name a occurring in $\mathcal{A}$, and an edge $\langle x_a, x_b \rangle$, with $\mathcal{L}(\langle x_a, x_b \rangle) = \{r \mid (a, b) : R \in \mathcal{A}\}$ for each pair (a, b) of individual names for which the set $\{R \mid (a, b) : R \in \mathcal{A}$ is non-empty.

The tableau algorithm applies the expansion rules presented in [2] where $R^{\mathcal{F}}(x, C) = \{y \mid y$ is R-successor of x and $C \in \mathcal{L}(y)\}$. The algorithm stops if it encounters a *clash*: a completion forest in which $\{A, \neg A\} \subseteq \mathcal{L}(x)$ for some node x and some concept name A or if there is some concept $\leq n \, R.C \in \mathcal{L}(x)$ and x has $n + 1$ R-successors $y_1, \ldots, y_n$ with $C \in \mathcal{L}(y_i)$ and $y_i \neq y_j$ for all $0 \leq i < j \leq n$. A completion forest is *clash-free* if none of its nodes contains a clash, and it is *closed* otherwise. It is complete if no rule can be applied to it. And the algorithm answers "$\mathcal{O}$ is inconsistent" if the completion forest contains a clash; and it answers "$\mathcal{O}$ is consistent" otherwise.

Note that the tableau algorithm for $\mathcal{ALCQ}$ ABoxes (i.e., TBoxes are empty) would always terminate. However, when the GCIs of TBoxes are discussed in the tableau algorithm, the algorithm might not be terminable. For instance, the algorithm for the GCI *Person* $\sqsubseteq$ ∃*HasParent.Person* runs perpetually. A so-called *blocking* technique is applied to guarantee termination of the expansion process even in the presence of GCIs. A node x is *blocked* if there is an ancestor y of x such that $\mathcal{L}(x) \subseteq \mathcal{L}(y)$ (called "y blocks x"), or if there is an ancestor z of x such that z is blocked; if a node x is blocked and none of its ancestors is blocked, then x is *directly blocked*.

We introduce a transformation $\sim$ defined as follows: (1) $\sim C(a) = \neg C(a)$; and (2) $\sim C \sqsubseteq D = C \sqcap \neg D(\iota)$ where ι is a special individual which does not occur before.

Lemma 1. *Let $\mathcal{O}$ be an ontology and ϕ a concept assertion or concept inclusion in $\mathcal{ALCQ}$. $\mathcal{O} \models \phi$ iff $\mathcal{F}$ is closed, where $\mathcal{F}$ is a complete forest of $\mathcal{O} \cup \{\sim \phi\}$ by applying the tableau algorithm.*

3 Forgetting in $\mathcal{ALCQ}$

In this section, following from forgetting $\mathcal{ALCQ}$ ontology presented in [3], we will simply give a semantic definition of what is means to forget about a set of variables in an $\mathcal{ALCQ}$ ontology.

As explained earlier, given an ontology $\mathcal{O}$ on signature $\mathcal{S}$ and $\mathcal{V} \subset \mathcal{S}$, in ontology engineering it is often desirable to obtain a new ontology $\mathcal{O}'$ on $\mathcal{S} - \mathcal{V}$ such that reasoning tasks on $\mathcal{S} - \mathcal{V}$ are still preserved in $\mathcal{O}'$. As a result, $\mathcal{O}'$ is weaker than $\mathcal{O}$ in general. This intuition is formalized in the following definition.

Definition 1. *Let $\mathcal{O}$ be an ontology in $\mathcal{ALCQ}$ and $\mathcal{V}$ a set of variables. An ontology $\mathcal{O}'$ over the signature $sig(\mathcal{O}) - \mathcal{V}$ is a result of forgetting about $\mathcal{V}$ in $\mathcal{O}$ if*

F1 $\mathcal{O} \models \mathcal{O}'$;
F2 *for each concept inclusion $C \sqsubseteq D$ in $\mathcal{ALCQ}$ not containing any variables in $\mathcal{V}$, $\mathcal{O} \models C \sqsubseteq D$ implies $\mathcal{O}' \models C \sqsubseteq D$;*
F3 *for each member assertion $C(a)$ or $R(a,b)$ in $\mathcal{ALCQ}$ not containing any variables in $\mathcal{V}$, $\mathcal{O} \models C(a)$ implies $\mathcal{O}' \models C(a)$ (resp., $\mathcal{O} \models R(a,b)$ implies $\mathcal{O}' \models R(a,b)$).*

If the result of forgetting about $\mathcal{V}$ in $\mathcal{O}$ is expressible as an $\mathcal{ALCQ}$ ontology, we say $\mathcal{V}$ is *forgettable* from $\mathcal{O}$.

Proposition 1. *Let $\mathcal{O}$ be an ontology in $\mathcal{ALCQ}$ and $\mathcal{V}$ a set of variables. If both $\mathcal{O}'$ and $\mathcal{O}''$ in $\mathcal{ALCQ}$ are resulting of forgetting about $\mathcal{V}$ in $\mathcal{O}$, then $\mathcal{O}' \equiv \mathcal{O}''$.*

This proposition says that the result of forgetting in $\mathcal{ALCQ}$ is unique up to ontology equivalence. Given this result, we write forget($\mathcal{O}, \mathcal{V}$) to denote any result of forgetting about $\mathcal{V}$ in $\mathcal{O}$ in $\mathcal{ALCQ}$. In particular, $\text{forget}(\mathcal{O}, \mathcal{V}) = \mathcal{O}'$ means that $\mathcal{O}'$ is a result of forgetting about $\mathcal{V}$ in $\mathcal{O}$.

If the result of forgetting about $\mathcal{V}$ in $\mathcal{O}$ is expressible as an $\mathcal{ALCQ}$ ontology, $\mathcal{V}$ is called *forgettable* from $\mathcal{O}$.

The following property states that the definition of the result of forgetting $\mathcal{ALCQ}$ ontology is appropriate.

Proposition 2. *Let $\mathcal{O}$ be an ontology and $\mathcal{V}$ a set of variables in $\mathcal{ALCQ}$. If both $\mathcal{O}'$ and $\mathcal{O}''$ are the result of forgetting about $\mathcal{V}$ in $\mathcal{O}$, then $\mathcal{O}' \equiv \mathcal{O}''$.*

Forgetting in TBoxes is independent of ABoxes as the next result shows.

Proposition 3. *Let $\mathcal{T}$ be a TBox in $\mathcal{ALCQ}$ and $\mathcal{V}$ a set of variables. Then, for any ABox $\mathcal{A}$ in $\mathcal{ALCQ}$, $\mathcal{T}'$ is the TBox of $\text{forget}((\mathcal{T}, \mathcal{A}), \mathcal{V})$ iff $\mathcal{T}'$ is the TBox of $\text{forget}((\mathcal{T}, \emptyset), \mathcal{V})$.*

Proposition 4. *Let $\mathcal{O}$ be an ontology in $\mathcal{ALCQ}$ and $\mathcal{V}$ a set of variables. Then*

1. *$\mathcal{O}$ is consistent iff $\text{forget}(\mathcal{O}, \mathcal{V})$ is consistent;*
2. *for any inclusion or assertion ϕ not containing variables in $\mathcal{V}$, $\mathcal{O} \models \phi$ iff $\text{forget}(\mathcal{O}, \mathcal{V}) \models \phi$.*

This proposition shows that two major reasoning tasks, namely, consistency and query answering, can be preserved in the definition of forgetting. From the property, such two reasoning tasks in an ontology can be reduced into those tasks in the result of forgetting in the ontology. In this sense, we take advantage of forgetting to optimize reasoning tasks.

The following proposition shows that the forgetting operation can be divided into steps, with a part of the signature forgotten in each step.

Proposition 5. *Let $\mathcal{O}$ be an ontology in $\mathcal{ALCQ}$ and $\mathcal{V}_1, \mathcal{V}_2$ two sets of variables. Then we have $\text{forget}(\mathcal{K}, \mathcal{V}_1 \cup \mathcal{V}_2) \equiv \text{forget}(\text{forget}(\mathcal{K}, \mathcal{V}_1), \mathcal{V}_2)$.*

For simplicity, forgetting in ontologies is independent of order of forgetting. Based on this idea, to compute the result of forgetting about $\mathcal{V}$ in $\mathcal{K}$, it is equivalent to forget in variables in $\mathcal{V}$ one by one.

4   Tableau-Based Forgetting in $\mathcal{ALCQ}$

In this section, we will compute the resulting of forgetting some variables based on the completion forest which is obtained by applying the tableau algorithm for $\mathcal{ALCQ}$.

Given an ontology $\mathcal{O}$ and a set of variables $\mathcal{V}$, the completion forest $\mathcal{F}$ which is obtained by applying the tableau algorithm may still contain some variables in $\mathcal{V}$. For instance, let $\mathcal{O} = (\{A \sqsubseteq B\}, \{A(a)\})$ and the completion forest $\mathcal{F}$ which is obtained by applying the tableau algorithm w.r.t. concept name A contains two branches $\mathcal{B}_1 = \{\mathcal{L}(a)\}$ where $\mathcal{L}(a) = \{A, \neg A\}$ and $\mathcal{B}_2 = \{\mathcal{L}(a)\}$ where $\mathcal{L}(a) = \{A, B\}$. However A still occur in $\mathcal{F}$. That is to say, in the completion forest $\mathcal{F}$, all variables forgotten are not deleted but ignored only. However, the result of forgetting does not contain any variable forgotten. Thus, to compute the result of forgetting from the $\mathcal{F}$, those variables forgotten are necessary to be deleted from $\mathcal{F}$. Since $\mathcal{F}$ are two different forms of the same result, we consider compute the result of forgetting based on $\mathcal{F}$ in this paper. In the following, we will delete variables in the completion forest by considering both nodes $\mathcal{L}(x)$ and edges $\mathcal{L}(\langle x, y \rangle)$ to a completion forest irrelevant to the variable set $\mathcal{V}$.

Definition 2 (Forgetting forest). *Let $\mathcal{O}$ be an ontology and $\mathcal{V}$ a set of variables. $\mathcal{F}$ is a completion forest by applying the tableau algorithm w.r.t. $\mathcal{V}$ on $\mathcal{O}$. We say the result of forgetting $\mathcal{V}$ in $\mathcal{F}$, written by* forget$(\mathcal{F}, \mathcal{V})$, *is a forest obtained by forgetting nodes (written* forget$(\mathcal{L}(x), \mathcal{V})$*) and forgetting edges (written* forget$(\mathcal{L}(\langle x, y \rangle), \mathcal{V})$*) defined as follows:*

- *for every node $\mathcal{L}(x)$,* forget$(\mathcal{L}(x), \mathcal{V})$ *is obtained from $\mathcal{L}(x)$ by*
  Step 1 *delete all the form $C \sqcup D$ or $C \sqcap D$ or $\exists R.C$ or $\geq nR.C$;*
  Step 2 *if $\{A, \neg A\} \subseteq \mathcal{L}(x)$ with $A \in \mathcal{V}$, then replace A and $\neg A$ by $\perp$;*
  Step 3 *if A or $\neg A$ or $A \sqcup C$ or $\neg A \sqcup C$ in $\mathcal{L}(x)$ with $A \in \mathcal{V}$, then delete A or $\neg A$ or $A \sqcup C$ or $\neg A \sqcup C$;*
  Step 4 *if $\forall R.C \in \mathcal{L}(x)$ or $\leq nR.C \in \mathcal{L}(x)$ with $R \in \mathcal{V}$, then delete $\forall R.C$ or $\leq nR.C$;*
  Step 5 *if $\forall R.C \in \mathcal{L}(x)$ or $\leq nR.C \in \mathcal{L}(x)$ with $R \notin \mathcal{V}$, then replace C with* forget$(\{C\}, \mathcal{V})$ *and delete $\forall R.(\top \sqcup C)$ or $\leq n \, R.C$;*
- *for every edge $\mathcal{L}(\langle x, y \rangle)$,* forget$(\mathcal{L}(\langle x, y \rangle), \mathcal{V})$ *is obtained from $\mathcal{L}(\langle x, y \rangle)$ by if $R \in \mathcal{L}(\langle x, y \rangle)$ with $R \in \mathcal{V}$, then $\mathcal{L}(\langle x, y \rangle) - \{R\}$.*

Note that (1) forget$(\mathcal{L}(x), \mathcal{V})$ is recursive; and (2) forget$(\mathcal{F}, \mathcal{V})$ is irrelevant to $\mathcal{V}$.

As will be readily seen, the forgetting forest algorithm w.r.t. nodes $\mathcal{L}(x)$ in completion forest $\mathcal{F}$ is similar to the algorithm of *compute C-forgetting* presented in [3]. It is quite natural that when we only consider each node $\mathcal{L}(x)$, the node $\mathcal{L}(x)$ can be taken as DNF of a *complex concept*. For instance, a node $\mathcal{L}(x) = \{A_1, A_2, \forall R.A_3\}$ can taken the DNF of the complex concept $C = A_1 \sqcap A_2 \sqcap \forall R.A_3$. We will apply the mechanism to compute the result of forgetting later. Forgetting forest algorithm w.r.t. edges $\mathcal{L}(\langle x, y \rangle)$ is directly deleting the roles in set of variables $\mathcal{V}$ from $\mathcal{L}(\langle x, y \rangle)$.

In fact, the forgetting forest algorithm holds the equivalence as follows.

Theorem 1. *Let $\mathcal{O}$ be an ontology and ϕ an axiom in $\mathcal{ALCQ}$. For any set of variables $\mathcal{V}$ irrelevant to ϕ, we have* forget$(\mathcal{O}, \mathcal{V}) \models \phi$ *iff* forget$(\mathcal{F}, \mathcal{V})$ *is closed. Where $\mathcal{F}$ is a completion forest of $\mathcal{O} \cup \{\sim \phi\}$ by applying the tableau algorithm.*

Given an ontology $\mathcal{O}$ and a set of variables $\mathcal{V}$, Theorem 1 shows that the forest which does not contain any variable in $\mathcal{V}$ obtained by applying the forgetting forest algorithm could capture the consistency of $\mathcal{O}$ limited in the set of variables $sig(\mathcal{O} - \mathcal{V})$.

Acknowledgments. This work is supported by the program of Applied Mathematics Discipline of Shanghai Polytechnic University (XXKPY1604) and the open funding project of Key Laboratory of Computer Network and Information Integration (Southeast University), Ministry of Education.

References

1. Baader, F., Calvanese, D., McGuinness, D.L., Nardi, D., Patel-Schneider, P.F.: The Description Logic Handbook: Theory, Implementation, and Applications. Cambridge University Press, Cambridge (2003)
2. Horrocks, I., Sattler, U.: Decidability of SHIQ with complex role inclusion axioms. Artif. Intell. **160**(1–2), 79–104 (2004)
3. Wang, Z., Wang, K., Topor, R., Pan, Z.J.: Forgetting for knowledge bases in DL-Lite. Ann. Math. Artif. Intell. **58**(1–2), 117–151 (2010)
4. Wang, Z., Wang, K., Topor, R., Zhang, X.: Tableau-based forgetting in ALC ontologies. In: Proceedings of ECAI 2010, pp. 47–52 (2010)
5. Zhang, X.: Forgetting for distance-based reasoning and repair in DL-Lite. Knowl. Based Syst. **107**, 246–260 (2016)
6. Zhang, X., Wang, K., Wang, Z., Ma, Y., Qi, G., Feng, Z.: A distance-based framework for inconsistency-tolerant reasoning and inconsistency measurement in DL-Lite. Int. J. Approx. Reason. (2016). http://dx.doi.org/10.1016/j.ijar.2016.08.003

Mining RDF Data for OWL2 RL Axioms

Yuanyuan Li, Huiying Li[✉], and Jing Shi

School of Computer Science and Engineering,
Southeast University, Nanjing, China
{220141547,101010166,220151530}@seu.edu.cn

Abstract. The large amounts of linked data are a valuable resource for the development of semantic applications. However, these applications often meet the challenges posed by flawed or incomplete schema, which would lead to the loss of meaningful facts. Association rule mining has been applied to learn many types of axioms. In this paper, we first use a statistical approach based on the association rule mining to enrich OWL ontologies. Then we propose some improvements according to this approach. Finally, we describe the quality of the acquired axioms by evaluations on DBpedia datasets.

Keywords: Linked data · RDF · OWL2 · Association rule mining

1 Introduction

Nowadays, semantic applications are emerging continually leading to a fast growing number of knowledge repositories on the web. Ontologies are an effective way to improve the quality of linked datasets but many datasets still lack the well-expressive schemas to infer potential information. In our work, we suggest the use of association rule mining methods for discovering ontological knowledge from linked data itself.

The structure of the paper is organized as follows: In Sect. 2, we give an overview of related works. In Sect. 3, we introduce OWL2 RL and Association Rule Mining. In Sect. 4, we describe the methods of getting axioms in OWL2 RL and propose an improvement. In Sect. 5, we describe experiment results learned from two versions of DBpedia dataset. Section 6 draws conclusions from our work and provides a future work.

2 Related Works

Several methods have been raised adapting machine learning methods. In [6], Vector Space Model was applied to recognize disjoint classes. Nebot and Berlanga [2] take advantage of the schema-level knowledge to generate transactions which will later satisfy traditional association rules algorithms. Lorey et al. [3] compare positive and negative association rules to existing schemas for indicating potential modeling errors. Particularly related to our approach is the recent work by Völker et al. Fleischhacker and Völker [4] presents a set of inductive methods to automatically enrich ontologies. Völker et al. [7] use SPARQL queries and mine axioms in the OWL2 EL.

© Springer Nature Singapore Pte Ltd. 2016
H. Chen et al. (Eds.): CCKS 2016, CCIS 650, pp. 117–123, 2016.
DOI: 10.1007/978-981-10-3168-7_12

3 Preliminaries

The OWL2 RL profile is aimed at applications that require scalable reasoning without sacrificing too much expressive power. Table 1 gives the syntax and semantics of OWL2 RL axioms. The concept of association rules has been widely studied in data mining. A lot of approaches can be applied. We choose the Apriori [1] algorithm.

Table 1. Axioms available in OWL2 RL.

Name	DL syntax	Semantics
SubClassOf	$C \subseteq D$	$\{x \in C^I \Rightarrow x \in D^I\}$
EquivalentClasses	$C \equiv D$	$\{x \in C^I \Rightarrow x \in D^I \wedge x \in D^I \Rightarrow x \in C^I\}$
DisjointClasses	$C \subseteq \neg D$	$\{x \in C^I \Rightarrow x \notin D^I\}$
SubObjectPropertyOf	$r \subseteq s$	$\{(x, y) \in r^I \Rightarrow (x, y) \in s^I\}$
EquivalentObjectProperties	$r \equiv s$	$\{(x, y) \in r^I \Rightarrow (x, y) \in s^I \wedge (x, y) \in s^I \Rightarrow (x, y) \in r^I\}$
DisjointObjectProperties	$r \subseteq \neg s$	$\{(x, y) \in r^I \Rightarrow (x, y) \notin s^I\}$
ObjectPropertyDomain	$\exists r.T \subseteq C$	$\{(x, y) \in r^I \wedge x \in C^I\}$
ObjectPropertyRange	$\exists r^-.T \subseteq C$	$\{(x, y) \in r^I \wedge y \in C^I\}$
TransitiveObjectProperty	$r \circ r$	$\{(x, y) \in r^I \wedge (y, z) \in r^I \Rightarrow (x, z) \in r^I\}$
InverseObjectPropertyOf	r^-	$\{(x, y) \in r^I \Rightarrow (y, x) \in r^I\}$
SymmetricObjectProperty	$Sym(r)$	$\{(x, y) \in r^I \Rightarrow (y, x) \in r^I\}$
AsymmetricObjectProperty	$Asy(r)$	$\{(x, y) \in r^I \Rightarrow (y, x) \notin r^I\}$
FunctionalObjectProperty	$T \subseteq (\leq 1\ r)$	$\{(x, y) \in r^I \wedge (x, z) \in r^I \Rightarrow y = z\}$
InverseFunctionalObjectProperty	$T \subseteq (\leq 1\ r^-)$	$\{(x, z) \in r^I \wedge (y, z) \in r^I \Rightarrow x = y\}$
IrreflexiveObjectProperty	$Irr(r)$	$\{\{(x, x) \mid x \in \Delta^I\} \cap r^I = \emptyset\}$
DataPropertyDomain	$\exists R.T \subseteq C$	$\{(x, y) \in R^I \wedge x \in C^I\}$

4 Mining RDF Data for OWL2 RL Axioms

In this paper, we use association rule mining to learn OWL axioms. We will first employ SPARQL query language to get ontology information. Afterwards, we translate them into suitable transaction tables. Finally, we execute Apriori algorithm to discover association rules which can be translated into OWL axioms eventually.

We illustrate the methods of obtaining axioms through an extracted dataset from DBpedia in Fig. 1. Table 2 (PP is for PopulatedPlace) is the transaction table for class axioms. Each class is labeled with a set of integer identifiers expressing if one instance belongs to this class (*1* for positive and *0* for negative). What' more, if instance i is not declared to be an instance of class C, we can have $i \in \neg C$. For property axioms, we take transitivity as an example. Transitivity means that if property r is transitive and the statements $a\ r\ x$ and $x\ r\ b$ exist, and $a\ r\ b$ must exist too. Hence, the item $r \circ r$ which means $a\ r\ x$ and $x\ r\ b$ for an arbitrary instance $x \in N_I$ is added. Each transaction in

```
Wave_Rock    rdf:type    dbo:PopulatedPlace .      Millennium_Final    dbo:previousEvent    Halloween_Havoc .
Wave_Rock    rdf:type    dbo:Place .               WCW_Mayhem          dbo:previousEvent    Millennium_Final .
Machakos     rdf:type    dbo:PopulatedPlace .      WCW_Mayhem          dbo:previousEvent    Halloween_Havoc .
Machakos     rdf:type    dbo:Place .               SupperBrawl         dbo:previousEvent    Souled_Out .
Dominica     rdf:type    dbo:PopulatedPlace .      SupperBrawl         dbo:previousEvent    Starrcade .
Dominica     rdf:type    dbo:Country .             Souled_Out          dbo:previousEvent    Starrcade .
Dominica     rdf:type    dbo:Place .               Arum_alpinum        dbo:genus            Arum .
Doxey        rdf:type    dbo:PopulatedPlace .      Kiang               dbo:genus            Equus_(genus) .
Doxey        rdf:type    dbo:Location .            Kiang               dbo:genus            Asinus .
Doxey        rdf:type    dbo:Place .               Asinus              dbo:genus            Equus_(genus) .
Awre         rdf:type    dbo:PopulatedPlace .      Arum                dbo:genus            Carl_Linnaeus .
Awre         rdf:type    dbo:Place .
```

Fig. 1. Triples extracted from DBpedia dataset 2015.

Table 2. Transaction tables for class axioms.

URI	PP	Country	Place	Location	¬PP	¬Country	¬Place	¬Location
Dominica	1	1	1	0	0	0	0	1
Odanad	0	1	1	0	1	0	0	1
Wave_Rock	1	0	1	0	0	1	0	1
Machakos	1	0	1	0	0	1	0	1
Doxey	1	0	1	1	0	1	0	0
Awre	1	0	1	0	0	1	0	1

Table 3. Transaction tables for transitivity axioms.

URIs	previousEvent	genus	genus o genus	previousEvent o previousEvent
(SuperBrawl, Starrcade)	1	0	0	1
(WCW_Mayhem, Halloween_Havoc)	1	0	0	1
(Asinus, Equus_(genus))	0	1	1	0
(Arum_alpinum, Carl_Linnaeus)	0	0	1	0
(SuperBrawl, tarrcade)	1	0	0	1

tables represents one possible pair of instances (a, b) and contains all possible $r \; o \; r$ and r. The transaction tables for transitivity are generated in Table 3.

In our experiment, we suppose the confidence threshold to be 0.8. From Table 2, we find the itemset {*PopulatedPlace*, ¬*Country*} reaches a support value of 4. And confidence value of rule *PopulatedPlace* $\Rightarrow$ ¬*Country* is 0.8. Likewise, the confidence value of rule *Country* $\Rightarrow$ ¬*PopulatedPlace* is 0.5. Hence, rule *PopulatedPlace* $\Rightarrow$ ¬*Country* can be mined, but *Country* $\Rightarrow$ ¬*PopulatedPlace* cannot. As we all known, the disjointness axioms are symmetrical. In our experiments, we get 1066 pairs of classes having the form of $A \Rightarrow \neg B$ but no form of $B \Rightarrow \neg A$. In addition, the confidence value of rule *Location* $\Rightarrow$ *Place* is 1.0 stating that *Location* is a subclass of *Place*. While rule *Place* $\Rightarrow$ ¬*Location* has the confidence value of 0.833. This leads to confliction too. We find 982 such contradictive rule pairs.

From the above analysis, we make a little adjustment to our method. Support $(A \cup B)$ means the number of instances both A and B have. In order to guarantee the symmetry, we choose the smaller one of support (A) and support (B). Three scenarios can be used to verify the rightness of the formula (1).

$$\text{confidence}(A \Rightarrow \neg B) = \text{confidence}(B \Rightarrow \neg A) = 1 - \frac{\text{support } (A \cup B)}{\min\{\text{support } (A), \text{support } (B)\}} \tag{1}$$

The first one is that class A and B are intersected depicted by Fig. 2(a). We can describe this scenario by example of two classes from DBpedia dataset 2015. Class *Automobile* has 8302 instances, while *MeanOfTransaportation* has 266 instances. They have 116 common instances. Thus, the confidence of *Automobile ⇒ ¬Mean OfTransaportation* is 0.986, and *MeanOfTransaportation ⇒ ¬Automobile* is 0.563. But in our method, the confidence values of both are 0.564. They are not disjoint. The second one is depicted by Fig. 2(b). Class *Place* has 10298 instance and *NaturalPlace* own 454 instances. All the instances belonging to *NaturalPlace* also belong to *Place*, which means *NaturalPlace* is the subclass of class *Place*. According to Apriori algorithm, the confidence value of *Place ⇒ ¬NaturalPlace* is 0.955. While in our method, confidence values of these two classes are both 0. Classes are not disjoint at all. The last one is that class A has no common instance with class B in Fig. 2(c). The overlapped instance number is 0 and the confidence values are 1.

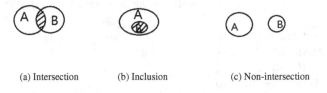

(a) Intersection   (b) Inclusion   (c) Non-intersection

Fig. 2. Relations between two classes

Finally, we present methods about getting property axioms in OWL2 RL.

Object Property Transitivity: From transaction Table 3. We can get the confidence of rule *previousEvent o previousEvent ⇒ previousEvent* 1. And the rule *genus o genus ⇒ genus* is 0.5. Thus, we can get the conclusion that object property *previousEvent* is transitive but object property *genus* is not.

Object Property Subsumption and Disjointness: These axioms are similar to the class axioms. Each transaction in the table represents one pair of instances (a, b) and contains all possible property items when tuple a r b holds in dataset. Association rule $r_i \Rightarrow r_j$ is used for subsumption. We extend the disjointness by adding $\neg r$ into itemset I just as classes. Rule $r_i \Rightarrow \neg r_j$ is for disjointness. In addition, the conflictions happened in class disjointness is also applied to property. Adjustment is applied too.

$$\text{confidence}(r_i \Rightarrow \neg r_j) = \text{confidence}(r_j \Rightarrow \neg r_i) = 1 - \frac{\text{support}\,(r_i \cup r_j)}{\min\{\text{support}\,(r_i), \text{support}\,(r_j)\}} \tag{2}$$

Other Properties: We have conducted other property axioms of OWL2 RL just like Fleischhacker [7] have already done.

5 Experiments

We run our experiment on two DBpedia datasets depicted in Table 4. All experiments have been conducted on a Windows system equipped with an Intel Xeon e3-1225 3.20 GHz processor and 16 G main memory. Three different confidence thresholds are applied to study the relationships between higher thresholds and the correctness of axioms. We set the support threshold to be 1.

Table 4. Statistical data from different version of DBpedia.

	DBpedia dataset 3.9	DBpedia dataset 2015
# of classes	434	677
# of object properties	685	671
# of data properties	689	686
# of instances	8432070	7204698

We mined 14 types of axioms for each dataset. Too many axioms are generated so that it is difficult to check the rightness of these axioms one by one. We randomly chose 50 axioms for each type. If less than 50 axioms, we chose all. The chosen axioms were evaluated by three ontology engineers in the form of a natural language sentence like "The domain of object property *starring* is the class *Film*". They had two choices *right* or *wrong* to evaluate. The accuracy of the learned axioms is computed by averaging the number of correctness from the three engineers. Table 5 gives the results.

According to results, we have some observations. It is noticeable that different confidence thresholds have little influence on the accuracy of our results. For the two DBpedia datasets, their numbers of each axiom are very similar except the domain and range axioms. That is in DBpedia dataset 2015, every property has at most one class as domain or range. While DBpedia 3.9 has more than one class as domain or range and these classes are equivalent or inclusive. What's more, low accuracy values for functional and inverse functional axioms come from an argument about the semantics, such as functional axioms for property color. One engineer thinks things may have at least one color while others think only one color is also ok sometimes.

Table 5. Evaluation with different confidence thresholds. Number of axioms annotated by #num and accuracy as Acc.

Axiom type	DBpedia dataset 3.9						DBpedia dataset 2015					
	0.8		0.9		1.0		0.8		0.9		1.0	
	#num	Acc	#num	Acc	#num	Acc	#num	Acc	#num	Acc	#num	Acc
$C_i \sqsubseteq D_j$	1930	0.95	1857	0.93	1527	0.95	1945	0.93	1941	0.92	1927	0.92
$C_i \sqsubseteq \neg D_j$	485414	0.90	485130	0.89	480671	0.93	185811	0.91	185736	0.93	184539	0.92
$r_i \sqsubseteq r_j$	45	0.96	40	0.95	33	0.94	46	0.89	39	0.91	29	0.93
$r_i \sqsubseteq \neg r_j$	448017	0.93	447868	0.93	445367	0.90	466941	0.92	466796	0.90	464271	0.92
$\exists r.T \sqsubseteq C$	419	0.88	336	0.90	100	0.92	3492	0.88	3368	0.82	2846	0.86
$\exists r^-.T \sqsubseteq C$	71	0.87	44	0.91	21	0.90	598	0.90	313	0.92	112	0.91
$T \sqsubseteq (\leq 1\ r)$	398	0.40	290	0.32	107	0.45	405	0.30	292	0.31	103	0.35
$T \sqsubseteq (\leq 1\ r^-)$	256	0.28	170	0.32	77	0.45	246	0.30	164	0.32	72	0.38
Sym(r)	4	1.0	2	1.0	0	0.0	2	1.0	0	0.0	0	0.0
Asy(r)	652	1.0	640	1.0	488	1.0	672	0.94	659	0.95	505	0.94
$r_i \sqsubseteq r_j^-$	14	0.29	10	0.4	8	0.5	7	0.43	3	1.0	1	0.0
$r \circ r \sqsubseteq r$	71	0.30	54	0.30	48	0.32	78	0.26	65	0.28	59	0.31
Irr(r)	670	0.98	669	1.0	572	1.0	685	1.0	683	0.98	540	0.97
$\exists R.T \sqsubseteq C$	186	0.88	146	0.90	58	0.92	2000	0.88	1944	0.88	1717	0.91

6 Conclusion and Outlook

In this paper, we mainly discussed the acquisition of various types of axioms from RDF data. We did experiments on different DBpedia datasets by means of association rule mining. After analyzing the acquired axioms, we found some deficiencies and proposed an improvement. Finally, the learned axioms were evaluated by three ontology engineers. In future, we will take other datasets into consideration such as Wikidata to improve the quality of axioms learning. New approaches should also be proposed to deal with constant updated datasets.

Acknowledgments. The work is supported by the Natural Science Foundation of Jiangsu Province under Grant BK20140643 and the National Natural Science Foundation of China under grant No. 61502095.

References

1. Agrawal, R., Srikant, R.: Fast algorithms for mining association rules. In: Proceedings of 20th International Conference on Very Large Data Bases, VLDB, vol. 1215, pp. 487–499 (1994)
2. Nebot, V., Berlanga, R.: Mining association rules from semantic web data. In: García-Pedrajas, N., Herrera, F., Fyfe, C., Benítez, J.M., Ali, M. (eds.) IEA/AIE 2010. LNCS (LNAI), vol. 6097, pp. 504–513. Springer, Heidelberg (2010). doi:10.1007/978-3-642-13025-0_52

3. Lorey, J., Abedjan, Z., Naumann, F., et al.: RDF ontology (re-)engineering through large-scale data mining. Semant. Web Chall. (2011)
4. Fleischhacker, D., Völker, J.: Inductive learning of disjointness axioms. In: Meersman, R., et al. (eds.) OTM 2011. LNCS, vol. 7045, pp. 680–697. Springer, Heidelberg (2011). doi:10.1007/978-3-642-25106-1_20
5. Völker, J., Niepert, M.: Statistical schema induction. In: Antoniou, G., Grobelnik, M., Simperl, E., Parsia, B., Plexousakis, D., Leenheer, P., Pan, J. (eds.) ESWC 2011. LNCS, vol. 6643, pp. 124–138. Springer, Heidelberg (2011). doi:10.1007/978-3-642-21034-1_9
6. Töpper, G., Knuth, M., Sack, H.: DBpedia ontology enrichment for inconsistency detection. In: International Conference on Semantic Systems, pp. 33–40. ACM (2012)
7. Fleischhacker, D., Völker, J., Stuckenschmidt, H.: Mining RDF data for property axioms. In: Meersman, R., et al. (eds.) OTM 2012. LNCS, vol. 7566, pp. 718–735. Springer, Heidelberg (2012). doi:10.1007/978-3-642-33615-7_18

A Mixed Method for Building the Uyghur and Chinese Domain Ontology

Yilahun Hankiz[1], Imam Seyyare[2(✉)], and Hamdulla Askar[1]

[1] Institute of Information Science and Engineering,
Xinjiang University, Urumqi 830046, China
8082453@qq.com, askarhamdulla@sina.com
[2] College of Politics and Public Administration,
Xinjiang University, Urumqi 830046, China
sayyarim@163.com

Abstract. The study of multilingual ontology on professional field is relatively rare, and a few of the many existing are about the public domain. This paper describes the mixed method for building a new multilingual ontology. By using the above mixed method, construct UC bilingual ontology about University Management field, through alignment and mapping the concepts and relations between the different language ontology then merging into one body - multilingual ontology. Finally, preliminary realized semantic query using SPARQL, so that can will provide basic support for minority languages cross-lingual retrieval from the perspective of the professional field.

Keywords: UC domain ontology · Semantic query · SPRQL · Cross-lingual retrieval

1 Introduction

When the World Wide Web has become the main source of knowledge for people, there are still some problems about low accuracy and low recall rate of information retrieval, even cannot searching any results information. Therefore how to obtain useful knowledge from massive information becomes an urgent problem to be solved. At the same time, the language using by the network is also more and more diverse. For retrieval problem, multi lingual feedback results are more comprehensive than monolingual feedback. Hence, people are no longer satisfied with the retrieval in the one language, instead they require to use a language to retrieve, and the results expressed by a variety of languages. Ontology as a model that can describe the relationship at the semantic level, it separates the structure and content of the information, and provides a clear representation of the semantic knowledge. So multi lingual ontology is the key to solve these problems [1]. Its key feature is corresponding concept's consistency in different language ontology. At present, most of the world's cross-lingual ontologies are based on the WordNet or using the same framework of WordNet's structure. For example, EuroWordNet, RussianWordNet, CCD and HowNet, and The Academia Sinica Bilingual Ontological WordNet (of China Taiwan), etc. [2]. The establishment of these multilingual ontology is a bridge for cross-language information processing. In digital

H. Chen et al. (Eds.): CCKS 2016, CCIS 650, pp. 124–129, 2016.
DOI: 10.1007/978-981-10-3168-7_13

library, the demand of multilingual information retrieval and mining is particularly significant [3]. However, in China, multilingual ontologies construction for Uyghur, Mongolian and Tibetan are still in its initial stage, in addition to Chinese, there are lack of or almost no other language's related research.

China is a unified multi-ethnic country, 53 of the 55 ethnic groups have their own language, which is closely related to the survival and development of the nation. Uyghur language is a mother language of the main ethnic minority (Uyghur) in Xinjiang and the surrounding areas. It is an adhesive language in morphological structure, and belongs to the Altai Turkic languages. There are vast and numerous classical literature, historical writings and translations in Uyghur language. Whether Uyghur language are as the main carrier of national culture heritage or as the main tool of spreading the knowledge of science and technology culture now, it is inestimable that the unique human culture value and the tremendous role in Xinjiang and its surrounding areas.

2 Related Work

The State Council issued "China's ethnic policy and the national common prosperity and development" in 2009. The white paper pointed out: "in order to make the minority people share in the fruits of the information age, the state has adopted various measures to promote the healthy development of the national minority language and writing standardization and information processing" [4]. It has been more than 20 years to study the information processing technology of Uyghur language. Although there are had been made great progress and achieved a lot of results all aspects, but still cannot keep up with the development speed of the information age. If Uyghur language cannot enter the information age, it will lose the basic functions of the language and culture of the carrier, and also will be mercilessly abandoned by this era. Therefore, Uyghur information processing is directly related to the fate of the character, and its significance is self-evident. Because ontology construction is based on the common knowledge that between man and man, man and machine, machine and machine. So, it is increasingly urgent that the construction of the Uyghur ontology in Knowledge Engineering, NLP and other Artificial Intelligence.

As a preliminary work, in [5] artificial constructed Uyghur Ontology with pro-tege4.3 about Mathematic and Information Science using domain ontology construction method. This result more comprehensively collected special domain concepts and more accurately described them from a professional point. And can say it basically filed the gap about Uyghur Ontology research, provide the basis about cross-lingual retrieval of Mathematics and Information Science as well. However, the number of concepts and individuals is very small, and the hierarchical relations between them are relatively simple, need to further extend and improve. In [6], proposed query expansion technology based on WordNet, that constructed Uyghur semantic dictionary automatically based on WordNet, and did a further query expansion using this dictionary. The method is relatively simple, universal property is good as well, but the noise ratio is relatively large so that cause not very high accuracy.

Sum up the rules of common and unique expressions about different things of each national language, it is necessary to find the similarities and differences between them. Therefore, multilingual ontology which unified standard and unified interface will provide an important foundation for the application of multi-national language intelligent information processing, and speed up its implementation.

3 UC Domain Ontology

As a research focus and application purpose of cross-lingual information processing, this paper proposed the necessity of UC (short for Uyghur & Chinese) bilingual domain ontology construction.

3.1 Method for Bilingual Ontology Construction

Multilingual ontology construction is divided into three methods. One is building a new ontology from scratch, second is multilingual ontology mapping, and the last is ontology translation or localization. Generally, first method need great workload, the last one has used by many organization already.

1. Construct New One From Scratch

In the absence of source and target language ontology, should learn two language ontology then mapping or translate. This method extracts single language ontology using English & Japanese dictionary with definition sentences, then alignment the different ontology under lexical layer [7].

2. Multi Lingual Ontology Mapping

Mapping between multilingual ontology when there are source and target language ontology here. E.g. as a open resources, WordNet used by many researcher created multilingual ontology through mapping. Chinese and English multilingual upper mapping research Including: create bilingual ontology by bilingual alignment using WordNet and HowNet [8].

3. Ontology Translation Or Localization

Translate source to target ontology thus obtain multilingual ontology when there are existing source language ontology and no target here [3].

In this paper select the domain of University management and construct the new UC domain ontology by mixed using the 1st and 3rd method. That is, construct source (Chinese) ontology first, then translate and mapping it into target ontology (Uyghur), finally merge them into one. The 3 steps are shown in Fig. 1.

Step1: Because there are so limited text of professional domain in Uygur and rich of the Chinese relatively. So first determine the concepts, hierarchical relations, individuals and properties from Chinese domain text, then build C-Ontology (short for Chinese Ontology) using Protege4.3, finally store it with OWL file.

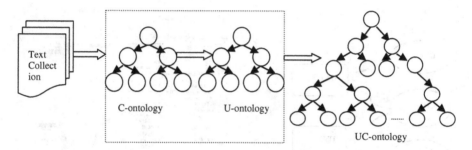

Fig. 1. Process of UC domain ontology construction

Step2: Then mapping the concepts and relations in C-Ontology into Uyghur concepts and relations. In the process, the lexical ambiguity is determined by the word attribute. ① To the nouns in the C-Ontology, first judge it whether single attribute or no. If it is not a single attribute and with the part of verb attribute from Chinese Uyghur Dictionary, then consider its noun only, and process the verb part as verb attribute. ② To the worlds with other attributes, process them use the same rule.③ In order to improve the accuracy of word matching, need to use the positive and negative matching strategy is needed. If there is one expression of the vocabulary that getting from Chinese Uyghur Dictionary, then use the only one directly. ④ If there are many expressions, then observe its word attribute. If it is single attribute then check up its reverse mapping. That is, if the mapping from Uyghur to Chinese by Uyghur Chinese Dictionary contain the source Chinese word then remain it or cancel it. ⑤ If the Chinese word attribute is not single then go ①. Figure 2 explains that, if want to get mapping of CH2 then remain UY1 only, because UY2's reverse mapping not contain CH2, so cancel UY2. Finally can build U-ontology (short for Uyghur ontology) matched with C-ontology using step1 after getting the concepts and properties through mapping.

Step3: Get the UC-ontology through merging the C-ontology and U-ontology with the same domain. The UC-ontology contain all concepts and relations of the two ontologies and they are matched perfectly each other.

Ontology merging is an effective way of ontology integration, and a kind of method to solve the ontology heterogeneity to realize the reuse and sharing of ontology resources. Ontology merging with same language is divided into two types. That is the merging of ontology with different domain and the merging of with same domain [9]. The relatively concepts and relations are can match each other, so realized it using the *Import* function of Protege4.3. That is first importing U-ontology into C-ontology then get the new bilingual ontology, and also call them ontology localization. Figure 3 is the interface of UC-ontology with Protege4.3. The relatively concepts matched each other with "same as" relation in the UC-ontology.

E.g. the instances"武汉大学"and "ۋۇخەن ئۇنۇۋېرسىتى" are with red restriction shown in Fig. 3.

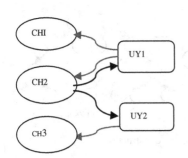

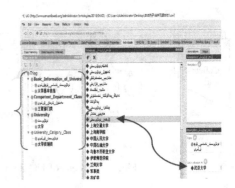

Fig. 2. Positive and negative matching strategy

Fig. 3. UC-ontology of university management domain

3.2   SPARQL Query on UC-Ontology

Jena is the Java framework of construction semantic web application program. It provide the best development environment for the ontology description language that OWL、RDF、RDFS etc. And it has the completely interface for function transfer and processing about ontology parsing, storing, reasoning and searching.

Now, as a kind of RDF Query Language, SPARQL (Simple Protocol and RDF Query Language standardized by the World Wide Web Consortium. Its importance is similar to SQL's for Relation Database. So it is the first choice of the query language for RDF, OWL etc. [10]. Therefore, in this paper select Jena as a the development environment of ontology and SPARQL as a ontology query language.

SPARQL allows query the triple in the ontology model with OWL file. Triple is similar to the "subject", "predicate" and "object" in natural language. Three tuple in ontology are <individual, property, value> and <class, property, value> etc. In this paper constructed the ontology with <individual, property, value>. When doing search on ontology, if know the one of the three tuple, then can search out all relatively three tuples. E.g. If we know the individual "شىنجياڭ ئۇنۇۋرستىي : ئۇرۇمچى" (Xinjiang University) in the ontology, then use the following SPARQL code can easily search out all of triple. That are classification of Xinjiang University, competent department, University place, University URL, ranking, grade, school motto, brief introduction, key lab account, and value of "same as", results shown in Fig. 4.

property	info
مەسئۇل_تارماق	ماڭارپ_نازارىتى
مەسئۇل_تارماق_قىسمى	"شىنجياڭ ئۇنىۋېر كايتچەنمۇ رايونلۇق ماڭارپ نازارىتى"^^xsd:string
قىسىم_تەرتىپى	"111"^^xsd:int
تۇرى	ئۇنىۋېرسال
مەسئۇل_تارماق_تۇرى	"شىنجياڭ ئۇرۇمچى شەھەرلىك تيانشان رايۇن غالبىيەت پولى229- نومۇر"^^xsd:string
ماكتەپ_تور_ئادرىسى	"www.xju.edu.cn"^^xsd:string
ماكتەپ_تەلسى	"لتىپاقلىق،ئەھرىبانە شاگىرتلاش،ئەسلىيەتچان،يېقىنلىق يارتىش"^^xsd:string
نوقتىلىق_تەجرىبىخانا_سانى	بەردەن:جەر نوقتىلىق ئالىي ماكتەپ دەپ بېكىتىلگ.1997-يىلى211-نۇرۇزۇلۇش ەدىكى نوقتىلىق ئالىي ماكتەپ قىلىپ بېكىتىلگەن" اماكتەپ_لسمۇچە_نۇنۇتۇرزۇنۇرزۇش
دەرىجىسى	"64"^^xsd:int
:sameAs	"ئالى يۇتنۇز"^^xsd:int
ئۇنىۋېرست_قورنى	untitled-ontology-22:新疆大学
	"ئۇرۇمچى شەەبى تيانشان رايۇنى غالبىيەت پولى14-نۇمۇر :"^^xsd:string

Fig. 4. SPAPRQL query results of UC-ontology triple

4   Conclusion

Ontology is an explicit formal specification of the domains and relations among them, and its goal is transforming the chaotic information into an orderly knowledge source for easy to use. This paper describes and designs the mixed method that building a new basic ontology from scratch, then get the multilingual ontology with university management domain through translation, mapping and merging. At last, implement the triple query of the multilingual ontology using SPARQL. From the perspective of professional domain, it will provide basic support for the cross-lingual information retrieval of minority languages. However, there are some problems, that Uyghur words are not very standard on user interface of Protege and Jena. Because Uyghur is agglutinative language, writing format differ from Chinese and English, and writing it from right to left, so there are some difficulties about the word processing. It is also the one of the next step to study and solve the problems of the paper.

Acknowledgments. This work was supported by the National Social Science Foundation of China (13BYY062).

References

1. Dai, W.: Technology and Method of Semantic Web Information Organization. Xue lin Press, Shanghai (2008)
2. Liu, Y., Lin, M.: Research on construction method of bilingual domain ontology based on OWL. Comput. Technol. Dev. **24**(8), 84–93 (2014)
3. Zhang, C.: Multilingual Domain Ontology Learning Research. Nanjing University Press, Nanjing (2012)
4. Zhao, X., Qiu, L., Zhao, T.: Construction technology of ontology knowledge base in minority languages. J. Chin. Inf. Process. **25**(4), 71–74 (2011)
5. Yilahun, H., Imam, S., Hamdulla, A.: A survey on uyghur ontology. Int. J. Database Theory Appl. **8**(4), 157–168 (2015)
6. Sawut, M.: Research on Query Extraction Technology Based on WordNet. Xinjiang University, Xinjiang (2015)
7. Nichols, E., Bond, F., Tanka, T., et al.: Multilingual ontology acquisition from multiple MRDs. In: Proceedings of the 2nd Workshop on Ontology Learning and Population (OLP2), Sydney, Australia, pp. 10–17 (2006)
8. Hu, H., Du, X.: Byuilding bilingual ontology from WordNet and Chinese classified thesaurus. In: Proceedings of the Scholl International Conference on Knowledge Science, Engineering and Management (KSEM2007), Melbourne, Australia, pp. 649–654 (2007)
9. Yi, L.: Imprecise Ontology Merging Research. Dalian Maritime University, Dalian (2010)
10. Ji, Z.: Ontology searching and reasoning. J. Micro Electron. Comput. **28**(10), 52–55 (2011)

Linked Data and Knowledge-Based Systems

Link Prediction via Mining Markov Logic Formulas to Improve Social Recommendation

Zhuoyu Wei[✉], Jun Zhao, Kang Liu, and Shizhu He

National Laboratory of Pattern Recognition, Institute of Automation,
Chinese Academy of Sciences, Beijing 100190, China
{zhuoyu.wei,jzhao,kliu,shizhu.he}@nlpr.ia.ac.cn

Abstract. Social networks have been a main way to obtain information in recent years, but the huge amount of information obstructs people from obtaining something that they are really interested in. Social recommendation system is introduced to solve this problem and brings a new challenge of predicting peoples preferences. In a graph view, social recommendation can be viewed as link prediction task on the social graph. Therefore, some link prediction technique can apply to social recommendation. In this paper, we propose a novel approach to bring logic formulas in social recommendation system and it can improve the accuracy of recommendations. This approach is made up of two parts: (1) It treats the whole social network with kinds of attributes as a semantic network, and finds frequent structures as logic formulas via random graph algorithms. (2) It builds a Markov Logic Network to model logic formulas, attaches weights to each of them to measure formulas contributions, and then learns the weights discriminatively from training data. In addition, the formulas with weights can be viewed as the reason why people should accept a specific recommendation, and supplying it for people may increase the probability of people accepting the recommendation. We carry out several experiments to explore and analyze the effects of various factors of our method on recommendation results, and get the final method to compare with baselines.

1 Introduction

Social networks have been a main way to obtain information in recent years. People get the latest news, knowledge of specific fields, or even just stories and jokes from them. There is a relationship between users called *Follow(follower, followee)*, which means the follower would like to pay attention to the followee or the content published by the followee. What kind of followees people followed determines what kind of messages they can get from social networks. Therefore, social recommend task can be view as predicting missing link on the social network.

An excellent social recommendation system can rescue people from searching and choosing, by bringing what they are interested in or helping them build new interests. At the beginning, the recommendation methods of e-commerce were

© Springer Nature Singapore Pte Ltd. 2016
H. Chen et al. (Eds.): CCKS 2016, CCIS 650, pp. 133–145, 2016.
DOI: 10.1007/978-981-10-3168-7_14

ported to social networks but the performance was not satisfactory. To improve the accuracy of recommendation, researchers propose a variety of solutions or techniques, such as taking explicit or implicit information into account, analyzing the user's social behaviors, and so on. They make the social recommendation techniques have a great development and go on improving apace. At the same time, many popular social networks, such as Twitter and Tencent Weibo, provided the reason why a user will accept a specific recommendation. For example, the reason can be that the user has 3 friends following the followee, or there is a high degree of similarity between tweets published by the user and the followee. Usually, labeling reasons to recommendations can increase the probability of users accept them.

However, there are still several problems which current recommended methods cannot solve well. The first and most headachy one is the cold start problem: Too little histories of new users make methods based on collaborative filtering failure. Secondly, heterogeneous attributes and relations cannot be modeled well nor used effectively. Although we have introduced a lot of features, such as users' age, tweets' keywords, social relations, location, accepted time, even the current mood of the users, few methods can effectively use them. Some methods assume all features are independent, which missed relevance between different types of relations and attributes. Others methods unite each two or more features to build new features and most of them are useless, which produce a huge feature space and make model extremely complex. Thirdly, the reasons why users will accept recommendations are generated from templates or rules listed manually, which takes a lot of time and may miss some cases.

In this paper, we propose a novel approach to bring logic formulas in social recommendation system and try to solve the above problems. Our method inherits the graph-based structure of the social works, and adds users' attributes to the graph. Distinguished from the Social-Attribute Network, the graph labels different semantic concepts to different types of nodes and edges. If we only take the concepts of nodes and edges into account, there are a lot of same structures (especially loops) on the graph. These conceptual structures can be viewed as frequent logic formulas from the perspective of first-order logic. We propose the Randomly Finding Loops algorithm to find these frequent logic formulas on the graph. Then we use the Markov Logic Network (MLN) to model directly logic formulas by treating each edges as a random variable and attach weights to formulas, rather than constructing, grounding a MLN and learning its structures and weights in the traditional way [24]. Finally, we construct queries with each user and followees recommended to it in the training data set, and learn the weights discriminatively.

We carried out several experiments on the Tencent Weibo data set and subsets from KDD-Cup'12 track 1 [21] to explore and analyze the effects of various factors of our approach on recommendation results, and then compare our approach with several baselines.

The major contributions of the paper are as follow: (1) We are the first to bring logic formulas in social recommendation system, and use them to represent the relations between social relations and kinds of attributes. (2) Distinguished from conventional methods based on random walk or grounding the MLN, we combine the advantages of both approaches by attaching weights to loops (formulas) to build the MLN directly, and learn the weights discriminatively rather than assigning value to them. (3) Our method generates reasons why users will accept recommendations automatically rather than manually.

The remainder of this paper is organized as follows. Section 2 introduces previous methods that are related to this work. Section 3 details how to use MLN to model social recommendation task. Section 4 details the Randomly Finding Loop Algorithm. Then experimental results are presented in Sect. 5, followed by the conclusion in Sect. 6.

2 Related Work

2.1 Social Recommendation Technique

Social recommendation techniques are different from traditional recommendation for e-commerce. It need model both uses' interests and items' (objects recommended) characters, and handle relations in social networks. Some traditional recommendation algorithms simply based on contents [3,4] or collaborative filtering [25] don't work well, because they cannot deal with heterogeneous data. At present there are already a number of models or methods who can handle such heterogeneous data containing attributes, relations and build a unify system to make recommendation. These methods can be divided into two categories, matrix factorization model [6,7] and graph-based model [2,8,27]. The former captures implicit relations between users and items, and merges all kinds of attributes,relations, even feedbacks [20] via factor vectors. The Matrix Factorization Model is the state-of-the-art method for collaborative filtering and collaborative ranking [16,17]. It uses factor vector to represent attributes, links, even users and items themselves, then the inner product of one user's vectors and one item's vectors is treated as final rating score. But it can only capture direct relations between factor vectors and create too many variables, which can lead to over-fitting and a long training time. Factorization machines [22,23] as an expansion of factorization model, it can handle more than two variables' interactions; While Karatzoglou et al. [12] solves the problem by expanding to the tensor decomposition approach. In this way, they must have more useless variables, which is a similar but more serious problem. While the latter transforms attributes to edges and combines into a heterogeneous graph, then applies random walking [2,5], propagation [13], paths finding, or just search techniques on it. Neighborhood-based methods are special cases of graph-based model when we only take 2-length paths into account. Item-based methods [25], user-based methods [26] and similarity calculating methods [9] all belong to neighborhood-based methods. For more general graph-based models: Social Attribute Network Model [8,28,29] creates an augmented network by adding attributes as nodes and undirect link between users and attributes; Methods based on propagations [11,30] defines a type of values and propagates them on the social relations graph. These methods can easily get useful multivariable interactions by randomly finding longer paths. However, they don't distinguish between types of paths and assign weights to paths directly accounting to degrees of nodes. We build a social semantic graph and use logic formulas to distinguish types and obtain weights from the learning process.

2.2 Markov Logic Network

Markov Logic Network (MLN) was first proposed by Pedro Domingos [24] formally. MLNs conbimes probability and logic by attaching weights to first-order formulas, and viewing these as templates for features of Markov Networks, and they can be applied to link prediction task. Recommendation systems can be views as practical instances of the the link prediction task. MLNs can find the relaiton paths called formulas. They are treated as a template of world together, and allocate weights attached to formulas to maximize the likelihood of the real world. Although MLNs can easily represent entities, attributes, and relationships in a social network, they rarely are applied to social recommendation system currently for its extreme computational complexity of learning or inference. Many techniques have been proposed to speed up this process: The discriminative learning methods [10,19] are used to decrease the number of random variables; Stanley Kok [14] clusters entities and relations before find formulas, and tries to find longer formulas by randomly finding motifs [15]. Though algorithm's efficiency has been improved, these methods still have to ground all relations with all entities. The grounding process spends a huge amount of time, even makes the problem cannot be computed. We abandon the grounding process and approximate the likelihood for learning by heuristic and stochastic sampling mechanism. This idea comes from finding frequent patterns algorithm on graph, such as Musk [1] and simpling DNF patterns [18], and it can be translate to learn structures of MLNs. This approach makes it possible to apply MLNs on large social recommendation data sets.

3   MLN for Social Recommendation

3.1 Building Social Semantic Graph

In the social network, we have a set of users noted as U and a set of attribute values of the users noted as A. In order to construct discriminative task, we create a subset of users noted as $U_i(U_i \subseteq U)$ for each User i in U as its alternative recommendations set, and User i can accept or ignore these recommendations. Then User i with each recommended user from U_i can be combined into a pair, noted as $Accept(user_i, user_r)$ in the form of triplets. They are called queries whose value is true when accepted or false when ignored. Our task is to predict the possibility of each query is true.

We build a direct semantic graph $G(N, E, C, R)$ whose nodes and edges are label types, to model U, A and all kinds of relations between them. N is the set of nodes and C is the set of nodes' types. All users in the social network and their attribute values are treated as nodes in N, and each nodes has a type in C, called *Concept*. E is the set of direct edges and R is the set of edges' types. All kinds of relations in social networks are treated as direct edges in E, and each edges also has a type in R, called *Relation*. The relation set contains social relations and action relations (e.g. *Retweet* action, *Comment* action and *At* action). From the perspective of Markov Networks, the triplets are treated as random variables and they are the nodes in the Markov Network. Therefore, we can introduce the MLN technology to create templates of the social semantic graph [24], which is the theoretical support of our approach.

In detail, the node set N should contain follow parts: (1) All users from U are added to N as nodes, and their concept is *user*; (2) All attribute values in A are treated as nodes with their attribute names as nodes' concepts. For example, *male* and *female* are nodes, whose concept is *gender*; Decades, such as 1990*s* and 2000*s*, are nodes, whose concept is *birthyear*; Keywords from users' statuses and comments are also nodes, whose concept is *keyword*.

Analogously, the edge set E contains follow parts: (1) The direct edge from *userA* to *userB* should be added, if userA has followed userB, noted as *Follow(userA, userB)*. (2) The own relations or mutex relations from U to A are added to E as edges, such as *Gender(user, male)*, *GenderFalse(user, female)*, *BirthYear(user, 1990s)*, *Keyword(user, keyword)* and so on.

3.2 Markov Logic Formula

If we want to estimate the possibility of that *Accept(user, user)* is true, we need treat it as a query, and then generate logic formulas containing the query. Meanwhile we need count the times of each logic formulas appearing. Here we show some examples for the logic formulas.

- $Follow(u_A, u_D) \wedge Follow(u_B, u_D) \wedge Follow(u_B, u_C) \Rightarrow Accept(u_A, u_C)$, $u_A \rightarrow u_D \leftarrow u_B \rightarrow u_C \leftarrow u_A$
- $Keyword(u_A, k_1) \wedge Keyword(u_C, k_1) \Rightarrow Accept(u_A, u_C)$, $u_A \rightarrow k_1 \leftarrow u_C \leftarrow u_A$
- $Keyword(u_A, k_2) \wedge Keyword(u_C, k_2) \Rightarrow Accept(u_A, u_C)$, $u_A \rightarrow k_2 \leftarrow u_C \leftarrow u_A$

All triplets on the left side of '$\Rightarrow$' are evidences of the query on the right side. We treat triplets as random variables, where the probability of evidences are true is 1 while the probability of query need to be estimated. Then we build a clique with all triplets in the same formula, and use the MLN to model it. The triplets are atomic, and we assum the atoms in evidence set are independent of the query. For MLNs, this means that the Markov Blanket of a query only contains evidence atoms [10].

From the perspective of the social semantic graph, these above are all entitative loops with u_A as start point and end point, examples of which are displayed behind the above logic formulas. Such entitative loops, or called entitative formulas, can be generated by running the finding loops algorithm on the social semantic graph, which will be detailed in Sect. 4. What's more, the processes of finding loops for different queries are independent of each other, so we parallelize these processes to make full use of computing resource.

Algorithm 1. Process Framework

PROCESS FRAMEWORK

1   **Build** static Social Attribute Graph
2   Start to **Maintain** global formulas set Υ
3   **for** each y in QuerySet **Y**
4   **FindLoop** for query y
5   **Count Locally** logic formulas
6   **BuildDataPoint** for query y
7   **Learn** weights for global formulas with DataPoints

Replace nodes in these entitative loops with their concepts, and we get conceptual loops. The formula set Υ are made up with these conceptual loops in MLNs [24], and a weight is attached to each of them. Finally, the weight vector **w** can be obtained from discriminative learning with train recommendations and it plays a decisive role in discriminating for test recommendations [10].

3.3   Discriminative Weight Learning

In this sub-section, we learn the weights of all conceptual formulas. We maximize the conditional log-likelihood (CLL) of the MLN with regularization, which is classic model of discriminative learning MLNs.

We create a query for each pair of a User u and a recommended User i from the alternative recommendations set U_i, noted as $Accept(u, i)$. We put all such queries into the query set Y, and run the finding loops algorithm for them. The conceptual formulas and counts obtained from the process are treated as features of the data point for the query y. In this way, we get the $|Y|$ data points as training data. Therefore, the CLL of Y is expressed as following under the evidence set X:

$$CLL = \sum_{k=1}^{n} \log P(Y_k = y_k | X = x) \tag{1}$$

Where k means the kth data point and Y_k is the kth query's label, whose value is 1 or 0 and noted as y_k, representing whether the recommendation is accepted. And,

$$P(Y_k = y_k | X = x) = \frac{e^{\sum_{j \in \Upsilon_{Y_k}} w_j n_j (x, y[Y_k=y_k])}}{e^{\sum_{j \in \Upsilon_{Y_k}} w_j n_j (x, y[Y_k=0])} + e^{\sum_{j \in \Upsilon_{Y_k}} w_j n_j (x, y[Y_k=1])}} \tag{2}$$

Where Υ_{Y_k} is the set of conceptual formulas with at least one entitative loop be found in finding loops for data point k. w_j is the weight of the jth formula, whose index j is global. $n_j(x, y[Y_k = y_k])$) is the number of the jth conceptual formula's true entitative loops, and similarly for $n_j(x, y[Y_k = 0])$) and $n_j(x, y[Y_k = 1])$).

Then reviewing the second and third logic formulas in Sub-sect. 3.2, we find the only difference between them is linked by different keywords. We need take the difference into account, because different entities brings different contributions. We assign a value for each entitative relation, and different types are calculated in different ways: (1) For *Follow* and *Accept*, their values are still 1; (2) For three action relations, *At, Retweet, Comment*, we normalize the counts of action relations (i.e. $At(user_A, user_B)$'s count) by the total action counts of the user, and take them as values for these entitative relations; (3) *Keyword* relations' values are set to their $tf - idf$ or other token-document values. After defining the values of edges, we use the following equation to calculate the value for a loop.

$$v(L) = \sqrt[n]{\prod_{i=1}^{n} v(E_i)} \tag{3}$$

Where n is the length of the loop. The equation eliminated the effect caused by different lengths of loops. And the $P(Y_k = y_k | X = x)$ Eq. (2) changes into:

$$P(Y_k = y_k | X = x) = \frac{e^{\sum_{j \in \Upsilon_{Y_k}} w_j V_j(x, y[Y_k = y_k])}}{e^{\sum_{j \in \Upsilon_{Y_k}} w_j V_j(x, y[Y_k = 0])} + e^{\sum_{j \in \Upsilon_{Y_k}} w_j V_j(x, y[Y_k = 1])}} \tag{4}$$

Where we use V_j to take the place of n_j, and V_j is the sum of the jth conceptual formula's entitative loops' values $v(L)$.

We take the negative CLL as the loss function and minimize it. Add the L2-regularization as an additional term, C as the regularization coefficient, and the loss function changes to $L(\mathbf{w}) = CLL + C\|\mathbf{w}\|_1$. The main process is sketched in Algorithm 1.

4   Randomly Finding Loops

4.1   Find Loops for a User

For a query, *Accept(user, recommend)*, we want to find formulas like this: $Relation_\pm$ (user, $node_1$) $\wedge$ $Relation_\pm$($node_1$, $node_2$) $\wedge$ ...... $\wedge$ $Relation_\pm$ ($node_{n-1}$, recommend) $\Rightarrow$ *Accept(user, recommend)*. Where $Relation_\pm$ ($node_1, node_2$) represents one of the two direct edges, $Relation(node_1, node_2)$ and $Relation(node_2, node_1)$. Remove edges' relations and represent the loop as a sequence of nodes, $user \rightarrow node_1 \rightarrow node2 \rightarrow ...... \rightarrow node_{n-1} \rightarrow recommend \rightarrow user$. Sequences of nodes like these can be got by searching and traversing on a simplified undirect graph, which was built by ignoring concepts of nodes, relations and directions of edges. Then for each two adjacent nodes in one sequence, we can get directly a entitative relation set of all $Relation_\pm(node_1,$

$node_2$) from the primary social semantic graph. Finally, a cartesian product of these relation sets of adjacent nodes can be treated as a set of all entitative formulas of the node sequence for the query.

The $Accept(user, recommend)$ as a query is always on the right side of $' \Rightarrow '$, and the evidences on the left side belong to the social semantic graph's edges set E which are all true. According to a first-order logic, the truth value of one formula is as same as the query's.

For the queries of one same User u and different recommended user in U_i, we make the following merger: (1) Change finding loops to finding paths; (2) All recommended users in User u's recommended user subset U_i are put into end node set EN; (3) Find paths staring with User u and ending with any recommended user in U_i, then make sure the lengths of these paths is limited between the maximum and the minimum. (4) Remove users never appearing as end nodes from U_i, because they are useless for model training.

We have to put forward a few rules, which can be view as pruning strategies: (1) prohibit backtrack to avoid getting palindromic node sequences, but the loop whose is 2 and the two edges are not identical are kept; (2) a loop containing a query edge as evidence is pruned, because queries have no exact truth values.

The time complexity of the algorithm for all users in U is $O(|U|H^L)$. Where H is the average size of each node's adjacent node set in N, and L is the maximum length we set for loops. Even when H and L are not very large, the H^L can be a enormous value and the time complexity is unbearable. Therefore, we have to give up complete search on the the whole social semantic graph.

4.2   Random Sampling

Retrospecting the Eq. (4), we want to estimate $P(Y_k = y_k|X = x)$ for Query y_k and it is a ratio. It relates to the ratio of $V_j(x, y[Y_k = y_k])$ and the sum of $V_j(x, y[Y_k = 1, 0])$, which increasing results in $P(Y_k = y_k|X = x)$ increasing. So we do not care about exact values of $V_j(x, y[Y_k = y_k, 1, 0])$ but the ratio.

The ratio of positive queries and negative queries is determinate when data set is determinate either for training or testing, or even alternative recommendation set in real world. Therefore, we can get all $P(Y_k = y_k|X = x)$ approximatively by ensuring fair treatments of positive and negative examples, and fair treatments of all concept formulas. The simplest way is to random walk on the Graph G and to allocate the same probability to adjacent nodes of one node in the transition probability matrix, but less formulas can be found or many conceptual formulas' counts is 0 when the percentage of sampling is very low. If we raise the percentage, the sampling mechanism will become insignificant. We need a heuristic strategy to find as many entitative formulas as possible when finding loops and keep all $P(Y_k = y_k|X = x)$ changeless at the same time. While the greedy strategy is a good choice. It means the algorithm will choose next nodes close to the target items. When we random walk and arrive at $node_u$ and want to transfer to its adjacent nodes, traverse each node noted as $node_v$ in $adj(node_u)$ and turn up the transition probability from $node_u$ to $node_v$ when there is a target user in $adj(node_v)$. Where $adj(node_x)$ is the set of nodes adjacent to $node_x$. The specific allocation method for the transition probability matrix P is as follows:

$$P(u, v) = \begin{cases} \dfrac{d(u, v)}{\sum_{x \in adj(u)} d(u, x)} & v \in adj(u), \\ 0 & v \notin adj(u). \end{cases} \quad (5)$$

$$d(u, v) = \begin{cases} 1 & v \in adj(u) \text{ and no target}, \\ \dfrac{c}{l_v} & v \in adj(u) \text{ and existing targets}, \\ 0 & v \notin adj(u). \end{cases} \quad (6)$$

Where $d(u, v)$ denotes the weight on the $Relation(u, v) \in E$, l_v is the number of nodes adjacent to $node_v$, and c is a constant which is larger and the algorithm tends to be more greedy. Such a transition matrix can ensure the proportionality of sampling while realizing the greedy strategy, which was proved in [1].

In this way, the time complexity of the randomly finding loops algorithm changes to $O(|U|M^L)$, where M is the maximum number of nodes to be visited from a visted node. And it does't require a large M, which is a tradeoff between computing time and the number of conceptual logic formulas found.

5   Experiments

5.1   Dataset and Evaluation Metrics

We choose an open data set, the Tencent Weibo Data Set (TWDS) from KDD-Cup'12 track 1 [21], for our experiments. There are kinds of attributes, relations, even circumstances in the TWDS, and then we select a few representative ones, including *Follow* relations, *At*, *Retweet*, *Comment* actions with times, keywords with weights, gender, and birth year. The alternative recommendations sets of the TWDS are also a bit special, all users in which are specific ones distinguished from ordinary friends in social networks, which can be celebrities, famous organizations, some well-known groups, or anything is public and famous [21]. Table 1 shows the statistics of these data sets. The recommended task is a classic ranking task, so apply the evaluation metrics of ranking to the recommended task is convictive. The Mean Average Precision (MAP) is a popular rank evaluation method to evaluate the proposed approach [31]. The KDD-Cup'12 track 1 use $AP@3$ as the final evaluation metric, and we expand it to $AP@n$ as our evaluation measures, where n is set to 1, 3, 5, 10.

Table 1. The statistics of datasets

Train size	Test size	Repetitive rate	Train accept rate	Test accept rate
1392872	1196410	22.9%	13.0%	11.0%

5.2 Method Comparison

In this sub-section, we compare our method with several baselines. The detailed implementations are listed below:

- *RandomGuess*: It exchanges positions of the recommended users randomly as the final result. Concretely, we exchange 1000 recommended user pair randomly for a specific user after reading the test data, which ensures that the output is completely random. If there are results of other methods worse than this, these methods are useless.
- *ItemBased−CF*: Item-based collaborative filtering. It calculates similarity for each pair of two items and recommends items to one user, which are similar to items followed by the user. It is a representative neighborhood-based method and it is easier to realize because the number of items is much smaller than the users'. The similar between two items $Sim(i, j)$ is calculated by Eq. (7), where $Follow_i$ means the follower set of Item i.

$$Sim(i, j) = \frac{Follower_i \cap Follower_j}{\sqrt{|Follower_i|} \cdot \sqrt{|Follower_j|}} \qquad (7)$$

- *MatrixFactorization*: Matrix Factorization Model. It is an excellent approach for recommendation systems, which captures implicit relations between users and items. It construct factor vectors for each user and item by decomposing the rating matrix. The factor vectors got can be use to predict the missing rating, which is used to recommend. For a pair of a user and a item, the rating r_{ui} is calculated by Eq. (8) [17].

$$r_{ui} = \mu + b_i + b_u + q_i^T p_u \qquad (8)$$

Where p_u and q_i are respectively the factor vector of User u and Item i. We learn it by minimizing the squared error function (9), where r_t is the true value from the rating matrix.

$$L(u, i) = \sum_{(u,i) \in K} (r_{ui} - r_t)^2 + \lambda(\|p_u\|^2 + \|q_i\|^2 + b_u^2 + b_i^2) \qquad (9)$$

5.3 Results

Table 2 shows the results of all methods, and we can obtain the following observations:

(1) Our method performs best on *AP@1*, which indicate the formula-based method is inclined to predict the top result.
(2) The performance of our method decreases with N (in *AP@N*) increases, which indicates our method is not good at predicting missing links without strong evidence.

(3) Matrix Factorization outperforms ours on $AP@3$, and it is the state-of-the-art for TWDS dataset. The winner of the KDD Cup 2012 developed its method based on Matrix Factorization. However, it does not outperform on the top place of the recommendation, which is the most import for almost all link prediction tasks. Therefore, it is necessary to merge formula-based method and matrix factorization method to achieve higher quality social recommendation.

Table 2. Results for different ML

Methods	AP@1	AP@3	AP@5	AP@10
Random guess	0.097	0.180	0.202	0.186
ItemBased-CF	0.207	0.326	0.311	0.284
Matrix factorization	0.220	0.366	0.327	0.298
Our method	0.277	0.353	0.315	0.268

6   Conclusion and Future Work

This paper treats social recommendation as a link prediction task on the social graph, and proposes a formula-based method to construct probabilistic formulas to predict potential links. Our method employs MLN to merge the force of various logic formulas and we conduct an experiment on a public social recommendation dataset in KDD Cup 2012. Our method achieve a good performance and perform best on precision at the top place of recommendation list.

In the future, we will explore the different effect of formula-based methods and matrix factorization, and try to merge them. To achieve higher quality social recommendation, we will also try to employ distributional representation methods, which are proved effective on the knowledge base and may be also good at social recommendation.

Acknowledgments. This work was supported by the Natural Science Foundation of China (No. 61533018), the National Basic Research Program of China (No. 2014CB340503) and the National Natural Science Foundation of China (No. 61272332 and 61602479). And this work was also supported by Google through focused research awards program.

References

1. Al Hasan, M., Zaki, M.J.: MUSK: uniform sampling of k maximal patterns. In: SDM, pp. 650–661 (2009)
2. Backstrom, L., Leskovec, J.: Supervised random walks: predicting and recommending links in social networks. In: WSDM, pp. 635–644. ACM (2011)
3. Balabanović, M., Shoham, Y.: Fab: content-based, collaborative recommendation. Commun. ACM **40**(3), 66–72 (1997)

4. Basu, C., Hirsh, H., Cohen, W., et al.: Recommendation as classification: using social and content-based information in recommendation. In: AAAI/IAAI, pp. 714–720 (1998)
5. Burda, Z., Duda, J., Luck, J., Waclaw, B.: Localization of the maximal entropy random walk. Phys. Rev. Lett. **102**(16), 160602 (2009)
6. Chen, K., Chen, T., Zheng, G., Jin, O., Yao, E., Yu, Y.: Collaborative personalized tweet recommendation. In: SIGIR, pp. 661–670. ACM (2012)
7. Chen, T., Tang, L., Liu, Q., Yang, D., Xie, S., Cao, X., Wu, C., Yao, E., Liu, Z., Jiang, Z., et al.: Combining factorization model and additive forest for collaborative followee recommendation. In: Proceedings of the KDD Cup 2012 Workshop (2008)
8. Gong, N.Z., Talwalkar, A., Mackey, L., Huang, L., Shin, E.C.R., Stefanov, E., Song, D., et al.: Jointly predicting links and inferring attributes using a social-attribute network (san). In: ACM Workshop on Social Network Mining and Analysis (SNA-KDD) (2012)
9. Hannon, J., Bennett, M., Smyth, B.: Recommending Twitter users to follow using content and collaborative filtering approaches. In: Proceedings of the Fourth ACM Conference on Recommender Systems, pp. 199–206. ACM (2010)
10. Huynh, T.N., Mooney, R.J.: Discriminative structure and parameter learning for Markov logic networks. In: ICML, pp. 416–423. ACM (2008)
11. Jamali, M., Ester, M.: A matrix factorization technique with trust propagation for recommendation in social networks. In: Proceedings of the Fourth ACM Conference on Recommender Systems, pp. 135–142. ACM (2010)
12. Karatzoglou, A., Amatriain, X., Baltrunas, L., Oliver, N.: Multiverse recommendation: n-dimensional tensor factorization for context-aware collaborative filtering. In: Proceedings of the Fourth ACM Conference on Recommender Systems, pp. 79–86. ACM (2010)
13. Kashima, H., Kato, T., Yamanishi, Y., Sugiyama, M., Tsuda, K.: Link propagation: a fast semi-supervised learning algorithm for link prediction. In: SDM, vol. 9, pp. 1099–1110. SIAM (2009)
14. Kok, S., Domingos, P.: Learning Markov logic network structure via hypergraph lifting. In: ICML, pp. 505–512. ACM (2009)
15. Kok, S., Domingos, P.: Learning Markov logic networks using structural motifs. In: ICML, pp. 551–558 (2010)
16. Koren, Y.: Factorization meets the neighborhood: a multifaceted collaborative filtering model. In: SIGKDD, pp. 426–434. ACM (2008)
17. Koren, Y., Bell, R., Volinsky, C.: Matrix factorization techniques for recommender systems. Computer **42**(8), 30–37 (2009)
18. Li, G., Zaki, M.J.: Sampling minimal frequent boolean (dnf) patterns. In: SIGKDD, pp. 87–95. ACM (2012)
19. Lowd, D., Domingos, P.: Efficient weight learning for Markov logic networks. In: Kok, J.N., Koronacki, J., Lopez de Mantaras, R., Matwin, S., Mladenič, D., Skowron, A. (eds.) PKDD 2007. LNCS (LNAI), vol. 4702, pp. 200–211. Springer, Heidelberg (2007). doi:10.1007/978-3-540-74976-9_21
20. Ma, H.: An experimental study on implicit social recommendation. In: SIGIR, pp. 73–82. ACM (2013)
21. Niu, Y., Wang, Y., Sun, G., Yue, A., Dalessandro, B., Perlich, C., Hamner, B.: The tencent dataset and KDD-Cup'12. In: KDD-Cup Workshop, vol. 2012 (2012)
22. Rendle, S.: Factorization machines. In: ICDM, pp. 995–1000. IEEE (2010)
23. Rendle, S., Gantner, Z., Freudenthaler, C., Schmidt-Thieme, L.: Fast context-aware recommendations with factorization machines. In: SIGIR, pp. 635–644. ACM (2011)

24. Richardson, M., Domingos, P.: Markov logic networks. Mach. Learn. **62**(1–2), 107–136 (2006)
25. Sarwar, B., Karypis, G., Konstan, J., Riedl, J.: Item-based collaborative filtering recommendation algorithms. In: WWW, pp. 285–295. ACM (2001)
26. Shi, Y., Larson, M., Hanjalic, A.: Exploiting user similarity based on rated-item pools for improved user-based collaborative filtering. In: Proceedings of the Third ACM Conference on Recommender Systems, pp. 125–132. ACM (2009)
27. Yang, S.-H., Long, B., Smola, A., Sadagopan, N., Zheng, Z., Zha, H.: Like like alike: joint friendship and interest propagation in social networks. In: WWW, pp. 537–546. ACM (2011)
28. Yin, Z., Gupta, M., Weninger, T., Han, J.: LINKREC: a unified framework for link recommendation with user attributes and graph structure. In: WWW, pp. 1211–1212. ACM (2010)
29. Yin, Z., Gupta, M., Weninger, T., Han, J.: A unified framework for link recommendation using random walks. In: International Conference on Advances in Social Networks Analysis and Mining (ASONAM), pp. 152–159. IEEE (2010)
30. Zhang, J., Wang, C., Yu, P.S., Wang, J.: Learning latent friendship propagation networks with interest awareness for link prediction. In: SIGIR, pp. 63–72. ACM (2013)
31. Zhu, M.: Recall, precision and average precision. Department of Statistics and Actuarial Science, University of Waterloo, Waterloo, February, 2004

Graph-Based Jointly Modeling Entity Detection and Linking in Domain-Specific Area

Jiangtao Zhang[1,2(✉)] and Juanzi Li[2]

[1] The 305th Hospital of Chinese People's Liberation Army, Beijing 100017, China
zhang-jt13@mails.tsinghua.edu.cn
[2] Department of Computer Science and Technology,
Tsinghua University, Beijing 100084, China
lijuanzi@tsinghua.edu.cn

Abstract. The current state-of-the-art Entity Detection and Linking (EDL) systems are geared towards general corpora and cannot be directly applied to the specific domain effectively due to the fact that texts in domain-specific area are often noisy and contain phrases with ambiguous meanings that easily could be recognized as entity mention by traditional EDL methods but actually should not be linked to real entities (i.e., False Entity mention (FEM)). Moreover, in most current EDL literatures, ED (Entity Detection) and EL (Entity Linking) are frequently treated as equally important but separate problems and typically performed in a pipeline architecture without considering the mutual dependency between these two tasks. Therefore, to rigorously address the domain-specific EDL problem, we propose an iterative graph-based algorithm to jointly model the ED and EL tasks in domain-specific area by capturing the local dependency of mention-to-entity and the global interdependency of entity-to-entity. We extensively evaluated the performance of proposed algorithm over a data set of real world movie comments, and the experimental results show that the proposed approach significantly outperforms the baselines and achieve 82.7% F1 score for ED and 89.0% linking accuracy for EL respectively.

Keywords: Entity detection and linking · False entity mention · Domain-specific entity linking · Joint model

1 Introduction

The problem of entity linking (EL), which involves linking extracted entity mentions to corresponding Knowledge Base (KB) entries is starting from [1, 7]. However, most of existing approaches [3,4,18] aim at the general KBs and cannot be directly used in the domain-specific corpora. With the increasing demand for constructing and populating domain-specific KBs, domain-specific EL techniques have been emerging as an effective way to manage and query information for specific fields. The difficulty of domain-specific EL is that the entity mentions in domain-specific area are often potentially highly ambiguous and various:

© Springer Nature Singapore Pte Ltd. 2016
H. Chen et al. (Eds.): CCKS 2016, CCIS 650, pp. 146–159, 2016.
DOI: 10.1007/978-981-10-3168-7_15

(1) the same mention may refer to several different entities; (2) some extracted mentions in the text are just normal phrases and should not be link to the entities (i.e., False Entity Mention(FEM)). (3) some common phrases could be real entity mentions in domain-specific corpora. Recently a few works [10,17] begin to explore domain-specific EL task but these works do not fully consider these issues mentioned above. Therefore we argue that domain-specific EL techniques deserve much deeper exploration.

Moreover, in most literatures, ED (Entity Detection) and EL (Entity Linking) are frequently treated as equally important but separate problems and typically performed in a pipeline architecture without considering the mutual dependency between these two tasks [9]. Therefore, in this paper, we propose a novel graph-based joint model combining ED and EL on a movie review corpora by overcoming the following challenges:

Poor Mention Boundaries: Although EL task can go wrong even when provided correct mentions, a large number of EL errors are caused by poor mention boundaries. Although the poor boundary problem is addressed as longest coverage matching in DBpedia spotlight and keyphrase extraction in Wikify, the boundary problem is especially severe in the domain-specific area. For the example shown in Fig. 1, both *"wall"* and *"wall street"* could be potentially linked to corresponding entities (movies) and it is difficult to determine which one is correct by traditional pipeline-based approaches [6,7,12,14], which just take extracted named entities as input of EL without considering the uncertainty and imperfection of the named entity extraction process. Intuitively, if we can leverage the feedback information from EL (outside knowledge information) to direct the process of ED, the issue of poor mention boundaries could be addressed. Recently, some works [2,11,15] perform ED and EL process jointly but their techniques do not take this issue into consideration and are best-suited for general KB instead of domain-specific KB.

False Entity Mention (FEM): Many previous proposed approaches assume that each entity mention extracted from a text *should* be linked to an entity

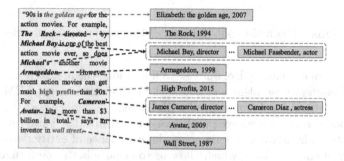

Fig. 1. An example for the task of domain-specific entity detection and linking

in the KB or NIL to indicate there is no matching entry [13]. However, in domain-specific area, such assumption may not be hold as in general corpora. For example, in Fig. 1, the extracted mentions *"wall street"*, *"golden age"* and *"high profits"* could be potentially linked to corresponding entities respectively because these mentions are the titles of entities representing movies. However, these mentions in this context are just common phrase and should not be identified as *true entity mention* (TEM). Therefore, we denote the extracted mention that *should not* be linked to any entites as the *False Entity Mention* (FEM). Notice that FEM is different with NIL (unlinkable mention) in some previous approaches [13] which indicates there is no matching entity in KB but should be linked.

Mutual Dependency: The main drawback of traditional pipeline-based approaches stems from the fact that they do not take into consideration the mutual dependency between ED and EL processes. But we argue that these two tasks are tightly coupled and the mutual information between these two tasks could be used to improve the performance of both. For example, in Fig. 1, the knowledge information of linking results of *"Avatar"* and *"The Rock"* could be helpful to filter out the FEM *"the golden age"* and *"wall street"* due to the fact that the main thread of this text is talking about the action movies but the movies *"the golden age"* and *"wall street"* are not action movies. Moreover, such information of ED results is also useful for the ranking of *"Michael"* and *"Cameron"*, which are the directors of movies *"The Rock"* and *"Avatar"* respectively.

Based on above observation, in this paper, we propose a new graph-based algorithm in specific domain via jointly incorporating ED with EL task. The main idea of our approach is as follows: First, we define and construct a Joint Graph based on work [5], our contribution is that the structure of our constructed graph encodes both the mention detection certainty, mention-to-entity linking confidence and the interdependent information between different entities together. Second, we calculate the initial score for each vertex and the weight of each edge in the graph. At last, we propose an iterative graph-based algorithm to step by step improve the detection accuracy and linking precision via propagating the interdependency between EL decisions.

Contributions. The main contributions of this paper are summarized as follows.

- To the best of our knowledge, our research is among the first to point the poor mention boundary problem and define the important concept FEM, both of which are critical for domain-specific EDL task.
- We proposed a novel iterative graph-based algorithm that jointly models ED and EL tasks by iteratively enhancing the confidence of entity detection and certainty of entity linking, which allow us to achieve better performance than the traditional EDL methods.

- To verify the effectiveness and efficiency of our proposed method, we conducted extensive experiments on a manually annotated dataset of real world movie comments and a domain-specific KB. The experimental results show the effectiveness of our proposed approach.

The remainder of this paper is organized as follows: Sect. 2 describes some preliminaries and give the task definition. Section 3 presents our Joint Graph for modeling ED and EL and Sect. 4 proposes the iterative algorithm based on constructed Joint Graph. Section 5 gives our experimental results, and Sect. 6 concludes.

2 Task Description

In this section, we begin by introducing some related concepts and notations. Next, we give the definition of our task.

2.1 Notations

Let $E = \{e_1, e_2, ..., e_{|E|}\}$ denotes the set of all entities of a domain-specific KB. Then we define a *mention* as a textual phrase (e.g., the *"the Rock"* in Fig. 1) which can *potentially* be linked to an entity in the domain-specific KB. Given a document d, we consider every possible n-gram (e.g. $n \leq 10$) as a *candidate mention* defined as $M = \{m_1, m_2, ..., m_{|M|}\}$. Further, we let $E(m_i) = e_{i1}, e_{i2}, ..., e_{i|E(m_i)|} \subseteq E$ denote the set of *candidate entities* which a *candidate mention* m_i could be linked to. For example, in Fig. 1, the set of entities that mention *"Cameron"* could be linked to is $E(\text{"Cameron"}) = \{\underline{\text{"James Cameron"}}, \text{"Cameron Diaz"}\}$. Specifically, we use $m.e \in E(m)$ to denote the true corresponding mapping entity of a mention m, i.g., the mapping entity of *"Cameron"* is *"James Cameron"*.

Notice that not all *candidate mentions* should be linked to entities. For example, *"wall street"* in the Fig. 1 is just a common textual phrase instead of a correct entity mention according to its context although there exists a movie named *Wall Street*. Therefore, we denote these mentions which *should not* be linked to any entities as *False Entity Mentions (FEMs)* which can be defined as $M_F = \{m_{f1}, m_{f2}, ..., m_{|M_F|}\} \subseteq M$. We also define those mentions that *should* be linked to entities in E as *True Entity Mentions(TEMs)* , which is denoted as $M_T = \{m_{t1}, m_{t2}, ..., m_{|M_T|}\} \subseteq M$. As the example shown in Fig. 1, the set of TEMs is $M_T = \{$ *"The Rock"*, *"Michael Bay"*, *"Michael"*, *"Armageddon"*, *"Cameron"*, *"Avatar"* $\}$ and the FEMs is $M_F = \{$ *"the golden age"*, *"high profits"*, *"wall street"* $\}$. Obviously, $M_T \cup M_F = M$.

2.2 Task Definition

The goal of our task is to map an extracted correct entity mention $m \in M$ to the corresponding entity $m.e \in E(m)$ in a domain-specific KB. In other words, given an input document d, we need to extract each TEM $m \in M_T$ and find its corresponding entity $m.e$ for each $m \in M_T$ while filtering out all FEMs from M. The input of our task is the text of a document d in a specific domain and a domain-specific KB pertaining to the domain while the output is the true entity mentions $M_T \subseteq M$ and their corresponding entities $\{m.e | \forall m \in M_T\}$.

Our task is composed by two joint parts, namely entity detection (ED) and entity linking (EL). ED is the task of identifying the boundaries, predicting the set M of given document d and extracting the true mentions $M_T \subseteq M$. EL is the task of disambiguating and linking each extracted mention $m \in M_T$ to its corresponding entity $m.e$ in the giving domain-specific KB.

3   The Joint Graph

3.1   Overview

In this subsection, we present the overview of our Joint Graph. Given a document d, We define the Joint Graph as $G = (V, A)$ where V is the vertexes set denoting all mention-to-entity pairs and E is the set of edges representing the interdependency between vertexes. Specifically, for each $m_i \in M$, and its candidate entity list $E(m_i) = \{e_{i1}, e_{i2}, ..., e_{i|E(m_i)|}\}$, the vertexes are formulated as a set $V = \{v_k = (m_i, e_{i,j}) | \forall e_{i,j} \in E(m_i), \forall m_i \in M, 1 \leq k \leq |V|\}$. Each vertex v_k in the graph is associated with an score $s(v_k)$ indicating the strength of detection certainty of m_i and linking confidence between m_i and $e_{i,j}$. For each pair of vertexes $\langle v_k, v_l \rangle$ in the graph, we add an undirected edge $\langle v_k, v_l \rangle$ to A, with a weight $w(\langle v_k, v_l \rangle)$ indicating the strength of interdependency between their entities. In this way, two types of dependencies are modeled in the Joint Graph:

1. Local dependency between mention and candidate entity
   In Joint Graph, the dependency between an entity mention m_i and a candidate entity $e_{i,j}$ is encoded as the score $s(v_k)$ of the vertex $v_k = (m_i, e_{i,j})$.
2. Global Interdependency between EL decisions
   By connecting candidate entities using the edges, the interdependency between EL decisions is encoded into the structure of the Joint Graph. In this way, the Joint graph allows us to deduce and use indirect and implicit dependency between different EL decisions. For example, the mention "*The Rock*" is related to the entity "*The Rock, 1994*", which in turn is related successively to the entity "*Michael Bay, director*". As a result, the relationship between "*Michael*" and "*Michael Bay, director*" could be strengthened while the relationship between "*Michael*" and "*Michael Fassbender, actor*" will be weakened.

For illustration, Fig. 2 shows the Joint Graph representation of the EDL problem in Example 1. To ease the representation, we do not draw all edges in the Joint Graph. From Fig. 2, we can see that the score of the TEMs vertex is high and there is a strong semantic relatedness between any two of the true mapping entities of TEMs. On the contrary, between the vertexes of TEM and FEM,

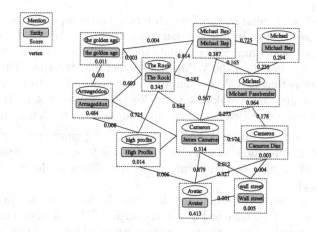

Fig. 2. The joint graph of example 1

the semantic relatedness is weak, which demonstrates that the Joint Graph can effectively model mention-to-entity linking confidence as vertex scores and entity-to-entity interdependency as edge weights.

3.2 Graph Construction

Before we construct the Joint Graph, we need to generate candidate mentions M and candidate entities $E(m)$ in the given document d first. Here, we consider every possible n-gram (e.g. $n \leq 10$) in d as a candidate mention and adopt the construction method described in [17] to overgenerate candidate mentions and entities.

The construction of Joint Graph takes two steps: vertexes generation and vertexes connection.

Vertexes Generation: Each mention $m_i \in M$ is paired with its every candidate entity $e_{i,j} \in E(m_i)$ in d to form a vertex in the Joint Graph. Then, each vertex $v_k = (m_i, e_{i,j})$ will be assigned to a score $s(v_k)$ to indicate the mention detection certainty of m_i and mention-to-entity linking confidence between $(m_i, e_{i,j})$, which will be introduced in Sect. 4.1.

Vertexes Connection: Next, we add the interdependent edge to the constructed vertexes. For each vertexes pair $\langle v_k, v_l \rangle$, $v_k = (m_i, e_{i,j})$, $v_l = (m_p, e_{p,q})$ in Joint Graph, if there is semantic relatedness between their entities (i.e. $e_{i,j}$ and $e_{p,q}$), we add an edge with weight $w(\langle v_k, v_l \rangle)$ between them to indicate their interdependent strength. Notice that Edges are not drawn between different nodes for the same mention since only one of candidate entities for the same mention may be the *true* mapping entity.

There has been several research which focused on computing the relatedness between entities [15,19]. In our approach, we adopt the Wikipedia Link-based Measure (WLM) algorithm [8] to calculate the relatedness of two entities $e_{i,j}$ and $e_{p,q}$. The WLM is based on the Wikipedia's hyperlink structure. The basic idea of this measure is that two Wikipedia articles are considered to be semantically related if there are many Wikipedia articles that link to both. We apply the same algorithm to our KB: Given two entity e_i and e_j, we define the semantic relatedness between them as $WLM(e_i, e_j) = 1 - \frac{\log(\max(|E_i|,|E_j|)) - \log(|E_i \cap E_j|)}{\log(|W|) - \log(\min(|E_i|,|E_j|))}$, where E_i and E_j are the sets of entities that link to e_i and e_j respectively in the KB, and W is the set of all entities in KB. Then we have $w(\langle v_k, v_l \rangle) = WLM(e_{i,j}, e_{p,q})$.

We show the example of semantic relatedness between vertexes in Fig. 2. The value shown beside each edge in Fig. 2 is the edge weight calculated using WLM. From Fig. 2, we can see that there is a strong relatedness relationship between any two of the *true* mapping entities.

4   Graph-Based Iteration Algorithm

4.1   Initial Score

In this section, we elaborate our iterative graph-based algorithm. First, each vertex $v_k = (m_i, e_{i,j})$ in the Joint Graph will be assigned with an initial score $s(v_k)$ indicating the confidence of a candidate mention being a TEM and the strength of a mention being linked to a candidate entity by leveraging the following four features.

Popularity: Most of current research [13,17], et al., use the popularity as an important feature in EL task which indicates popularity of a mention being linked to an entity by leveraging the count information from KB. Therefore, we formalize the popularity of a vertex $v_k = (m_i, e_{i,j})$ as follows:

$$pop(v_k) = \frac{count_{m_i}(e_{i,j})}{\sum_{e_{i,j} \in E(m_i)} count_{m_i}(e_{i,j})}, v_k = (m_i, e_{i,j}) \tag{1}$$

where $count_{m_i}(e_{i,j})$ is defined as the number of times that entity $e_{i,j} \in E(m_i)$ is linked by the mention m_i.

Linkable Probability: We also leverage the count information in the KB to get the linkable probability of a mention indicating the probability that a mention m_i is a TEM, which can be formalized as follows:

$$lp(v_k) = \frac{\sum_{e_{i,j} \in E(m_i)} count_{m_i}(e_{i,j})}{count(m_i)} \tag{2}$$

where $count_{m_i}(e_{i,j})$ is defined as the number of times that an entity $e_{i,j} \in E(m_i)$ is actually linked by the mention $m_i \in M$. $count(m_i)$ is the total number of appearances of mention m_i.

Coherence: One would expect that entities mentioned in the same context are likely to be topically coherent, i.e. they are likely semantic related [16]. Therefore, we exploit this semantic relatedness between entities in the document d to define the coherence feature $coh(v_k)$ of a vertex $v_k = (m_i, e_{i,j})$ as the average value of the semantic similarity between each context entity e_c and its entity $e_{i,j}$.

$$coh(v_k) = \frac{\sum_{e_c \in C_E(m_i)} SmtRel(e_c, e_{i,j})}{|C_E(m_i)|} \tag{3}$$

where $C_E(m_i)$ means the set of context entities which co-occur with m_i in the same document. In our algorithm, we also adopt WLM to get the semantic similarity $SmtRel(e_c, e_{i,j})$.

Context Similarity: It has been an effective way to use the context information to perform entity disambiguation. Therefore, we define the context similarity $c_s(v_k)$ of vertex $v_k = (m_i, e_{i,j})$ as the similarity between the context around m_i and the full text of $e_{i,j}$ via leveraging Jaccord algorithm.

$$cs(v_k) = Jaccard(S_m, S_e) = \frac{|S_m \cap S_e|}{|S_m \cup S_e|} \tag{4}$$

where S_m denotes the bag of words for context of m_i while S_e means the bag of words for the full text of $e_{i,j}$.

Based on these features illustrated above, we assign the initial score $s(v_k)$ for each vertex $v_k = (m_i, e_{i,j}) \in V$ as the weighted sum of these features as follows:

$$s(v_k) = \overrightarrow{W} \cdot \overrightarrow{F} \tag{5}$$

where $\overrightarrow{F} = \{pop(v_k), lp(v_k), coh(v_k), cs(v_k)\}$ is a feature vector, and $\overrightarrow{W} = \{w_1, w_2, w_3, w_4\}$ is a weight vector, $\sum w_i = 1$. The weight vector $\overrightarrow{W}$ can be easily learned by supervised machine learning technique such as SVM on a training data set. Obviously, the score of a vertex $v_k = (m_i, e_{i,j})$ indicates the certainty of m_i being a TEM and confidence of m_i being linked to $e_{i,j}$.

4.2   Iterative Algorithm

In order to simplify the description of our proposed iterative graph-based algorithm, we first introduce the following three notations for our graph-based algorithm:

S: The initial score vector $S = \{s_1, s_2, ..., s_{|V|}\}$, where $s_k = s(v_k)$.

S_f: The final score vector $S_f = \{s_{f1}, s_{f2}, ..., s_{f|V|}\}$, where $s_{fk} = s_f(v_k)$. To ease the presentation, we denote the final score vector S_f exactly after round r iteration as S_f^r.

B: we define the adjacency matrix of the Joint Graph G as the iteration matrix B. B is a $|V| \times |V|$ matrix, where the value of element $B[k, l]$ is the edge weight between vertex v_k and v_l.

To compute the final score vector S_f, we first set its initial value s_f^0 as the initial score vector S, i.e., $S_f^0 = S$. Then we can update the final score vector S_f in an iteration manner as follows,

$$S_f^{r+1} = \lambda S + (1 - \lambda) B S_f^r \tag{6}$$

where $\lambda \in [0, 1]$ is the relative importance fraction of the two parts, of which appropriate value will be evaluated in Sect. 5. From this equation, we can see that our algorithm combines information from the initial score vector S and the interdependent information between vertexes by updating the final score vector iteratively until the final score stabilizes within a certain iteration steps which is set to 10 in our experiment.

At last, we can choose the mapping entity $m_i.e$ for entity mention m_i as:

$$m_i.e = \arg\max_{e_{i,j} \in E(m_i)} s_f(v_k), v_k = (m_i, e_{i,j}), \forall e_{i,j} \in E(m_i) \tag{7}$$

Since there are FEMs in the given document, we have to deal with this problem by validating whether the returned entity $m_i.e$ with highest score according to Eq. 7 is a correct mapping entity for mention m_i. We adopt a simple method: learning a FEM threshold τ to validate the highest score entity. If the final score $s_f(m_i.e)$ is greater than the FEM threshold τ, we return $m_i.e$ as the correct mapping entity for entity mention m_i, otherwise we return it as FEM and treat it as common phrase. The FEM threshold τ is learned by linear search based on the training data set, which is set to 0.25 in our experiment.

5   Experiments and Evaluation

To evaluate the effectiveness and efficiency of our proposed approach, we present an extensive experimental study in this section. All the programs were implemented in Python and all the experiments were conducted on a server (with four 2.7 GHz CPU cores, 1024 GB memory, Ubuntu 13.10).

Data Set. We conduct experiments on a gold standard data set for our task and adopt the Keg-Movie-Ontology (KMO) as the target domain-specific KB which have been used in [17]. The KMO, constructed by knowledge engineering laboratory of Tsinghua University, is a high quality KB, which integrates several English and Chinese movie data sources from LinkedIMDB, Douban and Baidu Baike, and contains 23 oncepts, 91 properties, more than 700,000 entities and 10 million triples. The gold standard data set contains user comments from several well established websites in China, such as 163, sina, sohu and tianya, etc. which have been manually annotated.

Table 1 lists some statistical data of the gold standard data set. From the table we can see that there are 842 comments, which include 2529 FEMs and 11848 TEMs. The number of all candidate entities is 42105. Average number of mentions (includes TEMs and FEMs) in one comment and candidate entities per candidate mention is 17.05 and 2.92, respectively.

Table 1. Statistical data of the user data set

Documents	$|FEMs|$	$|TEMs|$	CEs	$|M|$	$|E(m)|$
843	2529	11848	42105	17.05	2.92

Baseline Methods. Due to the fact that the traditional approaches could not directly apply on our data set and KMO, we created two classic baselines employed the traditional pipeline architecture that takes extracted entity mentions as the input to the following EL task. Moreover, in order to fairly evaluate the effectiveness of our proposed approach, we also adopt the method used in [17], named IJM(Interactive Joint Model) as another baseline.

Prior Probability-based method (POP). In this baseline, we only use linkable probability and popularity for ED and EL respectively. We set a threshold and only retain the mention whose linkable probability is higher than the threshold which is set to 0.045 in our experiment. Then we choose the the entity with the highest popularity among all the candidate entities as the mapping entity for this entity mention.

Context Similarity-based method: ($CSim$). We constructed a context vector for each mention and a profile vector for each candidate entity (e.g. using TFIDF). Then we measure the similarity of these two vectors for each pair of a mention and a candidate entity (e.g. cosine distance). Finally, the entity with the highest similarity is considered as the mapping entity for the mention. We also set a threshed and only retain the mention whose highest similarity score is larger than the pre-set threshold which is set to 0.087.

Interactive Joint Model: (IJM). The method IJM, proposed in [17], used an interactive framework between ED and EL tasks to improve the performance of both tasks iteratively via updating the values of features of these two tasks in an interactive manner.

Evaluation Metrics. Our task involves jointly modeling ED and EL processes which influence each other, therefore we also adopt the evaluation metrics used in [17], i.e.,

- **ED:** precision, recall and F1-measure;
- **EL:** accuracy over correctly recognized entities;
- **Overall ED+EL:** precision, recall and f-measure; the precision/recall is computed as the product of the NER precision/recall by the EL accuracy.

Influence of Fraction Factor. λ the $\lambda \in [0, 1]$ is the relative fraction factor between the initial score and score of last iteration. From Eq. 6 we can see that if $\lambda = 0$, the iteration only considers interdependency propagation. If $\lambda = 1$, there is no iteration and only the initial score works. Thus, the value of λ indicates

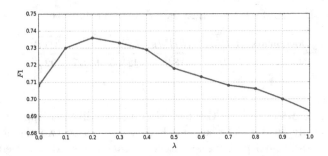

Fig. 3. F1 versus λ

the balance between the local dependency of mention-to-entity and global inter-dependency of entity-to-entity. We evaluate the relationship between the value of λ and the overall F1 score, as indicated in Fig. 3. From the figure we can see that when $\lambda = 0.2$, the F1 get the highest score. Therefore, in our experiment, the value of λ is set to the 0.2.

Result and Analysis. In order to evaluate the effectiveness of our jointly iterative graph-based algorithm, we configured the proposed approach into four different settings:

- *Fixed Weights+No Iteration (FW+NoI):* We don't use the machine learning method to train the weights. We assume that all features have the same weight, that is, the weight of all features is 0.25. Furthermore, we don't perform the iteration, i.e., $\lambda = 1$.
- *Initial Score+No Iteration (IS+NoI):* We use the initial scores computed by the Eq. 5 without performing the iteration, i.e., $\lambda = 1$ in Eq. 6.
- *Random Initial Score+Iteration (RIS+I):* We use random initial scores instead of the initial scores computed by Eq. 5 and perform the iteration according to Eq. 6.
- *Initial Score + Iteration (IS+I):* We use the initial scores computed by the Eq. 5 and perform the iteration according to Eq. 6.

Table 2 gives the comparison of our proposed approach and all other methods mentioned above. The experimental results demonstrate that different configurations of our proposed graph-based algorithm significantly outperforms the two baseline methods (i.e., *POP* and *CSim*) and our final approach *IS+I* also outperforms the *IJM* proposed in [17], which demonstrates the effectiveness of our proposed model.

In general, we can see that our proposed algorithm achieves high accuracy for EL in all configurations, which shows that our algorithm is very effective for EL task. The interdependency between the referent entities in the same document can provide critical evidence to the EL decision.

For the assessment of the *POP* baseline, obviously, the probability of being a TEM is high for the mention with high linkable probability. However, due

Table 2. Comparison of experiment results

Approach	Overall ED + EL			EL	ED		
	Precision	Recall	F1	Accuracy	Precision	Recall	F1
POP	0.615	0.509	0.557	0.792	0.776	0.643	0.703
CSim	0.597	0.590	0.594	0.825	0.724	0.715	0.719
IJM	0.741	0.690	0.714	0.875	0.847	0.788	0.816
FW+NoI	0.665	0.648	0.656	0.851	0.781	0.762	0.771
IS+NoI	0.717	0.670	0.693	0.866	0.828	0.774	0.800
RIS+I	0.727	0.660	0.692	0.859	0.846	0.768	0.805
IS+I	**0.764**	**0.710**	**0.736**	**0.890**	**0.858**	**0.798**	**0.827**

to *POP* uses the method of simply setting a threshold to exclude the mention with small linkable probability, *POP* gets a high precision but low recall. For the *CSim* baseline, because it considers context rather than prior probability, the recall of *CSim* is higher than *POP*, but the precision of *CSim* is damaged because it also introduces the FEMs.

Additionally, for different configurations of our algorithm, the performance of *FW+NoI* improves both ED and EL performance than baselines because four features are considered not merely prior probability. The performance of *IS+NoI* further improves as it considers the importance of different features by leveraging machine learning techniques. Meanwhile, the key point of the *RIS+I* is to investigate the influence of the iteration without considering the initial scores. The results indicate that overall precision further improves due to the fact that iteration exclude FEMs effectively while recall falls because no feature is considered.

Moreover, although *IJM* consider the interaction of ED and EL and use an interactive framework to jointly model these two tasks, our proposed method *IS+I* outperforms the *IJM* due to the fact that it decodes both the local dependency of mention-to-entity and global interdependency of entity-to-entity into a joint graph and use a similarity-flooding-like algorithm to propagate the dependency.

Finally, as expected, by modeling and exploiting local dependency of mention-to-entity and global interdependency of entity-to-entity, the final configuration of our method *IS+I* gets the highest performance in terms of overall precision and recall which achieved 32 % F1 improvement compared with the baseline *POP*, 24 % F1 improvement compared with the baseline *CSim* and 3 % F1 improvement compared with *IJM*.

6   Conclusion

The traditional EDL systems aim at general domain area. An unfortunate effect of this aim is that such generalist systems are often disappoint when they are applied to domain-specific area. Furthermore, most of existing EDL techniques ignore examining the interdependency of entities extraction and linking. In this paper, we proposed and evaluated an iteratively joint graph-based algorithm to model the ED and EL task by capturing the local dependency of mention-to-entity and global interdependency of entity-to-entity. The experiment results show that our proposed approach offers competitive performance to the three baseline systems, which indicate that it will be very useful for the domain-specific applications.

Acknowledgments. The work is supported by 973 Program (No. 2014CB340504), NSFC-ANR (No. 61261130588), Tsinghua University Initiative Scientific Research Program (No.20131089256), Science and Technology Support Program (No. 2014BAK04B00), and THU-NUS NExT Co-Lab.

References

1. Bunescu, R., Pasca, M.: Using encyclopedic knowledge for named entity disambiguation. In: Proceedings of the 11th Conference of the European Chapter of the Association for Computational Linguistics (EACL2006), pp. 9–16 (2006)
2. Guo, S., Chang, M.W., Kiciman, E.: To link or not to link? A study on end-to-end tweet entity linking. In: HLT-NAACL, pp. 1020–1030 (2013)
3. Han, X., Sun, L.: A generative entity-mention model for linking entities with knowledge base. In: Proceedings of the 49th Annual Meeting of the Association for Computational Linguistics: Human Language Technologies - Volume 1, pp. 945–954 (2011)
4. Han, X., Sun, L.: An entity-topic model for entity linking. In: EMNLP-CoNLL 2012, pp. 105–115 (2012)
5. Han, X., Sun, L., Zhao, J.: Collective entity linking in web text: a graph-based method. In: Proceedings of the 34th International ACM SIGIR Conference on Research and Development in Information Retrieval, pp. 765–774 (2011)
6. Lin, T., Mausam, E., O.: Entity linking at web scale. In: AKBC-WEKEX 2012, pp. 84–88 (2012)
7. Mihalcea, R., Csomai, A.: Wikify!: linking documents to encyclopedic knowledge. In: Proceedings of the Sixteenth ACM Conference on Conference on Information and Knowledge Management, pp. 233–242 (2007)
8. Milne, D., Witten, I.H.: Learning to link with wikipedia. In: Proceedings of the 17th ACM Conference on Information and Knowledge Management, pp. 509–518 (2008)
9. Nguyen, D., Theobald, M., Weikum, G.: J-NERD: joint named entity recognition and disambiguation with rich linguistic features. Trans. Assoc. Comput. Linguist. **4**, 215–229 (2016)
10. Olieman, A., Kamps, J., Marx, M., Nusselder, A.: A hybrid approach to domain-specific entity linking. CoRR (2015)
11. Pu, K.Q., Hassanzadeh, O., Drake, R., Miller, R.J.: Online annotation of text streams with structured entities. In: CIKM, pp. 29–38 (2010)
12. Ratinov, L., Roth, D., Downey, D., Anderson, M.: Local and global algorithms for disambiguation to wikipedia. In: HLT, pp. 1375–1384 (2011)
13. Shen, W., Wang, J., Jiawei, H.: Entity linking with a knowledge base: issues, techniques, and solutions. IEEE Trans. Knowl. Data Eng. **27**, 443–460 (2014)

14. Sil, A., Cronin, E., Nie, P., Yang, Y., Popescu, A.M., Yates, A.: Linking named entities to any database. In: EMNLP-CoNLL, pp. 116–127 (2012)
15. Sil, A., Yates, A.: Re-ranking for joint named-entity recognition and linking. In: CIKM, pp. 2369–2374 (2013)
16. Sil, A., Yates, A.: Re-ranking for joint named-entity recognition and linking. In: Proceedings of the 22nd ACM International Conference on Information Knowledge Management, pp. 2369–2374 (2013)
17. Zhang, J., Li, J., Li, X.L., Shi, Y., Li, J., Wang, Z.: Domain-specific entity linking via fake named entity detection. In: DASFAA, pp. 101–116 (2016)
18. Zhang, W., Sim, Y.C., Su, J., Tan, C.L.: Entity linking with effective acronym expansion, instance selection and topic modeling. In: IJCAI 2011, pp. 1909–1914 (2011)
19. Zhang, W., Su, J., Tan, C.L., Wang, W.T.: Entity linking leveraging: automatically generated annotation. In: Proceedings of the 23rd International Conference on Computational Linguistics, pp. 1290–1298 (2010)

LD2LD: Integrating, Enriching and Republishing Library Data as Linked Data

Qingliang Miao[1(✉)], Ruiyu Fang[1], Lu Fang[1], Yao Meng[1],
Chenying Li[2], Mingjie Han[2], and Yong Zhao[2]

[1] Fujitsu R&D Center Co. Ltd., Chaoyang District,
Beijing 100027, People's Republic of China
{qingliang.miao,fangruiyu,
fanglu,mengyao}@cn.fujitsu.com
[2] China Agricultural University, Haidian District, Beijing 100083,
People's Republic of China
{licy,hanmj,zhaoyong}@cau.edu.cn

Abstract. The development of digital library increases the need of integrating, enriching and republishing library data as Linked Data. Linked library data could provide high quality and more tailored service for library management agencies as well as for the public. However, even though there are many data sets containing metadata about publications and researchers, it is cumbersome to integrate and analyze them, since the collection is still a manual process and the sources are not connected to each other upfront. In this paper, we present an approach for integrating, enriching and republishing library data as Linked Data. In particular, we first adopt duplication detection and disambiguation techniques to reconcile researcher data, and then we connect researcher data with publication data such as papers, patents and monograph using entity linking methods. After that, we use simple reasoning to predict missing values and enrich the library data with external data. Finally, we republish the integrated and enriched library data as Linked Data.

1 Introduction

Libraries are experiencing a time of huge, tumultuous change. With the rapid development of digital libraries, library management agencies and users are faced with an increasing amount of publications. The huge amount and not interconnected nature of publications challenges library management agencies and users on managing and accessing scientific information. On the one hand, users demand intelligent search services to discover interested publications. On the other hand, library management agencies need to incorporate semantic information to better organize their digital assets and make publications more discoverable. For example, many libraries maintain data on researchers, papers and other materials, and separate search systems are built for each of these data sets [1]. As data is so distributed and heterogeneous, there is not a single search engine that can effectively retrieve a comprehensive set of the resources, e.g. find all the papers related to a given author within a given time period. Libraries have all been exploring new approaches to dramatically improve the discovery experience for

H. Chen et al. (Eds.): CCKS 2016, CCIS 650, pp. 160–171, 2016.
DOI: 10.1007/978-981-10-3168-7_16

users seeking scholarly information resources, such as traditional monograph and journal publications, archival materials, web archives, and much more [2].

Moreover, researchers are duplicated and ambiguous. One researcher may have different mentions (names) that distributed in different data sets, while different researchers may have the same name. Therefore, disambiguating and detecting duplicated researchers are necessary. If we can detect duplicated and ambiguous researchers, library management agencies and users can use the library data more efficiently. Since library data covers many elements such as papers, patents, discipline and organizations, it contains a large ration of missing values in its data sets. The impact of missing values is even aggravated when combining different data sets. The missing values makes library data harder to integrate and link. Consequently, missing value complement and data enrichment are important.

The Semantic Web in general and the Linked Data[1] initiative in particular encourage institutions to publish, share and interlink their data. This has considerable potential for libraries, which can complement their data by linking it to other external data sources. The Linked Data technology meets the need of connecting distributed data silos across the web. The Linked Data is based on a set of principles created by W3C[2]. The primary data model of Linked Data is the Resource Description Framework (RDF)[3], under which each resource in Linked Data space is identified by a unique HTTP dereferenceable Uniform Resource Identifier (URI) and the relations of resources are described with simple subject-predicate-object triples. Based on these principles, resources are linked by relations, and sophisticated networks of Linked Data can be built.

In this paper, we present the first effort to work on integrating, enriching and republishing library data as Linked Data. More specifically, we adopt Linked Data technology to integrate library data that wasn't previously linked. We first use hierarchical clustering method to conduct duplicated detection and disambiguation for researchers. And then, we link researchers with other library data such as monograph, journal publications, archival materials, research results, images and recordings. After that, we enrich library data by predicting missing values and republish library data as Linked Data. Our contributions are:

– We analyze and integrate several data sources including library data, DBpedia, Zhishi.me.
– We provide a system architecture for transforming library data into Linked Data including data cleaning, data integration, data enrichment and republishing.
– We use reasoning method to predict missing values and enrich the library data with external data.
– We develop a system[4] providing semantic search, statistical analysis and visualization based on linked library data.

---

[1] http://linkeddata.org/home.
[2] http://www.w3.org/.
[3] http://www.w3.org/RDF/.
[4] http://36.110.45.42:3333/.

The remainder of the paper is organized as follows. In the next section we review the related literature on linked library data. In the third section, we introduce the Chinese Agriculture University (CAU) Library data. We introduce the approach in detail and present the results in the fourth section. Last, we conclude the paper with a summary of our work and point out future directions.

2   Related Work

There are three related research field to our work. They are person disambiguation, entity linking and property alignment in the following subsections respectively.

2.1   Person Disambiguation

Previous work usually uses clustering techniques to solve person disambiguation issues. Monz and Weerkamp [3] introduce a clustering approach to person name disambiguation. Yoshida et al. [4] propose to use a two-stage clustering algorithm by bootstrapping to improve person disambiguation performance, and they use named entities, compound key words, and URLs as features for similarity calculation. Xu et al. [5] present a new key-phrased clustering method combined with a classification to improve cluster performance. Cucerzan [6] proposes a name entity disambiguation method through a process of maximizing the agreement between the contextual information extracted from Wikipedia and the context of a document, as well as the agreement among the category tags associated with the candidate entities. More recently, researchers combine traditional disambiguation methods with Linked Data knowledge for entity disambiguation. For example, Damljanovic and Bontcheva [7] combine a state-of-the-art entity disambiguation tool with novel Linked Data-based similarity measures and show that the combined algorithm can improve disambiguation accuracy. Usbeck et al. [8] propose a novel knowledge-base-agnostic approach for named entity disambiguation. Their approach combines the Hypertext-Induced Topic Search (HITS) algorithm with label expansion strategies and string similarity measures.

2.2   Entity Linking

Entity linking has attracted more and more attentions from both academia and industry. For example, Mihalcea and Csomai [9] propose Wikify system to annotate text using Wikipedia. Milne and Witten [10] implement a similar system called Wikipedia Miner, which adopts supervised disambiguation approach using Wikipedia hyperlinks as training data. Han and Sun [11] leverage entity popularity and context knowledge for entity linking. In practical applications, TagMe [12] system adopts a collective disambiguation approach, which computes agreement score of all possible bindings, and uses heuristics to select best target. DBpedia Spotlight [13] is a system for automatically annotating text with DBpedia. One important feature of the system is that it allows users to configure the annotations through the DBpedia ontology and quality measures such as prominence, topical pertinence, contextual ambiguity and disambiguation confidence.

The disambiguation model of Illinois Wikifier [14] is based on weighted sum of features such as textual similarity and link structure. AIDA [15] is a robust system based on collective disambiguation exploiting the prominence of entities, context similarity between the mention and its candidates, and the coherence among candidate entities for all mentions.

2.3 Property Alignment

Since different data sets may use different properties, property alignment should be conducted. Property alignment is related to schema matching and ontology matching. Falcon-AO [16], Logmap [17] RiMOM [18], and PARIS [19] are ontology matching tools for the automatic alignment of instances, properties and classes from different ontologies. These tools reach satisfactory results in the recent OAEI evaluation. Different from traditional ontology alignment settings, in this study, domains and ranges of properties are not provided. Worse still, some object values are missing. Lack of such ontological knowledge, these tools fail to conduct property alignments.

3 Data Sources

In this study, we use CAU library data. The CAU library data set contains data ranging from 1980 to 2015 and it contains 108340 entities in 10 isolated data sets. The statistics of CAU library data is shown in Table 1. Our goal is to integrate these 10 isolated data sets, enrich these data semantically, and republish them as Linked Data.

Table 1. The statistics of CAU library data

Data set	#Instance	#Property
Researcher	5863	51
SCI indexed journal paper	11934	42
Chinese journal paper	48449	52
Thesis	32755	41
Patent	2389	24
Project	3941	31
Monograph	1572	61
Research results	536	30
Curriculum	410	42
Organization	312	46
Discipline	179	52

CAU library data has a large proportion of missing values. Due to the page limit, we only shows the statistics of instances missing discipline and affiliation values in Table 2.

Table 2. The statistics of missing value in CAU library data

Data set	# Instance	#NoDiscipline	#NoAffiliation
SCI indexed journal paper	11934	48	959
Chinese journal paper	48449	27104	12407
Thesis	32755	673	5691
Patent	2389	1	2389
Project	3941	1677	2751
Monograph	1572	1150	1572
Research results	536	3	536
Curriculum	410	10	55

In this study, we use simple reasoning method to predict missing values as detailed in Sect. 4.5. Besides missing values, we enrich CAU library data by linking it with external knowledge base e.g. DBpedia [20] and Zhishi.me [21] as well.

DBpedia, initially released in 2007, is an effort to extract structured data from Wikipedia and publish the data as Linked Data. Zhishi.me is the first effort to publish large scale Chinese semantic data and link them together as a Chinese LOD (CLOD). Zhishi.me derives important structural features in three largest Chinese encyclopedia sites (i.e., Baidu Baike, Hudong Baike, and Chinese Wikipedia) and proposes several data-level mapping strategies for automatic link discovery. At present, the CLOD has more than 5 million distinct entities.

DBpedia and Zhishi.me could supply more information for instances in CAU library data. For example, when linking researcher with DBpedia and Zhishi.me, more information can be obtained such as nationality, birthday, birthplace, research field and awards. When linking organization instance with DBpedia and Zhishi.me entity, more information can be obtained, such as past name, launch date, longitude, latitude, homepage. Moreover, linking research topic with DBpedia and Zhishi.me entity, we can obtain category information by "dc:subject" relation and other mentions by DBpedia redirection relation.

4   The Approach

In this section, we will first illustrate the system architecture of the proposed approach, and then introduce how to integrate and link these data silos into Linked Data, and how to enrich the Linked Data with external knowledge base.

4.1   System Architecture

Figure 1 shows the system architecture of the proposed approach. The inputs are structured data in CSV or XML format and unstructured text and html data, and the outputs are linked library data. The approach includes five main modules: (1) duplication detection and disambiguation; (2) data linkage; (3) ontology design; (4) data enrichment and (5) data republish.

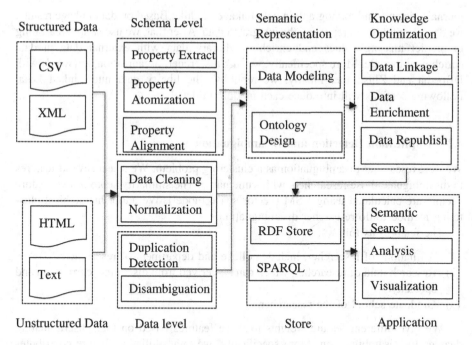

Fig. 1. LD2LD system architecture

Firstly, the input data is preprocessed at both schema and data level. Schema lever preprocessing includes property extraction, atomization and alignment. Some properties in original data is non-atomized, for example, some properties indicate time period information, therefore, they should be separated into two properties indicating starting date and ending date respectively. Some properties including time modifier, e.g. "2010 PhD entrance examination subjects" should be separated as well. Data level preprocessing includes data cleaning and normalization. For example, there are more than 30 different time expressions in CAU library data, therefore, we prepare specific normalization rules for each time expression. Besides time expression, we develop normalization rules for currency as well. If a string value contains any delimiter, the value is segmented into different parts by the delimiter and each segment will be assigned a type. For example, "BEIJING AGR UNIV, COLL ANIM SCI & TECHNOL, BEIJING 100094, PEOPLES R CHINA" will be segmented as "BEIJING AGR UNIV", "COLL ANIM SCI & TECHNOL", "BEIJING 100094", "PEOPLES R CHINA" and assigned types "University", "College", "Address", "Country". Since CAU library data was created and managed by different agents, they may use different properties to represent the same thing. For example, property "学科专业" is used in SCI Indexed Journal Paper data set, while "相关一级学科" and "相关二级学科" are used in Chinese Journal Paper data set. Therefore, we need to conduct property alignment.

After preprocessing the input data, we conduct duplication detection and disambiguation for researchers and assign a URI for each researcher. This URI is essential for the integration and enables to link researcher with other publication data such as journals, patents and monograph in data linkage module; Based on data linkage results, we design ontology to represent the integrated data. After that, we use simple reasoning to predict missing values and enrich the library data with external data in data enrichment module. More specifically, we link researcher, organization, keywords with DBpedia and Zhishi.me. Finally, we republish the library data into Linked Data. Following sections will introduce each module in detail.

4.2 Duplication Detection and Disambiguation

We treat researcher disambiguation as a clustering problem. We select several features to disambiguate those researchers with same name. The similarity score of two feature vectors are calculated using VSM (vector space model). We adopt hierarchical clustering method to do researcher disambiguation.

The features as follows:

(1) Affiliations of researcher, include college and department of researcher.
(2) Research field of researcher, which can be derived from discipline, curriculum and specialty.
(3) Graduate school of the researcher.

We give different feature weights to those features based on their discriminating degrees for disambiguation. More specifically, we treat affiliation feature contributes more to disambiguate two researchers with same name than other types of feature. If several same named researchers hold the same affiliations, we prefer identifying them as the same person. And features are combined using pre-defined weights, we try different groups of feature weights and select the one with best performance. Given two feature vectors $P_1 = (a_{11}, a_{12}, \ldots, a_{1n})$ and $P_2 = (a_{21}, a_{22}, \ldots, a_{2n})$, a_{ij} takes the value of 0 or 1, which stands for whether the feature condition is met. Meanwhile we define a group of feature weights $W = (w_1, w_2, \ldots, w_n)$, $\sum_{i=1}^{n} w_i = 1$. And the similarity score of two researcher vectors is computed using formula (1):

$$sim(P_1, P_2) = \sum_i w_i a_{1i} a_{2i} \tag{1}$$

During hierarchical clustering, to decide which clusters should be combined, we adopt the average linkage criterion as in formula (2).

$$csim(c_1, c_2) = \frac{1}{|c_1||c_2|} \sum_{p_i \in c_1} \sum_{p_j \in c_2} sim(p_i, p_j) \tag{2}$$

There are 5863 researchers in original CAU library data, after disambiguation and duplicated detection, we get 5583 researchers. We find 297 different researchers with same name and 130 duplicated records of 65 researchers. We conduct a preliminary experiment to evaluate the duplicated detection and disambiguation results, and the accuracy of hierarchical clustering method is 98%.

4.3 Data Linkage

After researcher disambiguation and duplicated detection, we link researchers to their archive i.e. SCI indexed journal papers, Chinese journal papers, theses, monographs, curriculums, patents, projects and research results. Data linkage, however, can be no-trivial due to the researcher ambiguity and name variation issues. The researcher ambiguity issue means that a mention could refer to multiple researchers in different data sets. Name variation indicates that an entity may be mentioned in different ways such as official name, nickname, aliases, abbreviation or even misspellings. For example, researcher names of SCI papers are usually written in abbreviated form. Therefore, cross-lingual data linkage is more complicated due to the cross-lingual ambiguity. To solve these issues, we extract rich features from both researcher profiles and their archive, and compute the similarity of two feature sets, and link two resources if their similarity score is greater than a threshold.

Since SCI papers are written in English, meanwhile the researcher profiles are in Chinese form. To solve the cross-lingual linking issues, we develop a cascaded linking method. More specifically, we first link resources (researcher profiles and archive) in the same language. Then we enrich the researcher profile feature sets by adding new features extracted from the linking results obtained in the first step. As a result, researcher profile feature set is enriched. Then we translate the enriched feature set into English: 1. we translate the coauthor names into English, in both complement and abbreviation forms. 2. We translate the publication titles and keywords into English. After feature set translation, we conduct mono-lingual linking using the method described above. We also use a self-training strategy by iteratively adding confident features into the researcher feature sets during linking.

To evaluate the data linkage performance, we manually annotate 10 researchers and their archive as the test data. Table 3 shows the experiment results.

Table 3. The experiment result of data linkage

Data set	Precision	Recall	F1-measure
SCI indexed journal paper	0.989	1.0	0.994
Chinese journal paper	1.0	1.0	1.0
Thesis	0.994	1.0	0.997
Patent	1.0	1.0	1.0
Project	1.0	1.0	1.0
Research results	1.0	1.0	1.0

4.4 Ontology Design

Selecting established ontologies as the basis for data modeling is strongly suggested in the semantic web community, since it makes the published data easier to share and exchange. Consequently, we aimed to do that as well. In practice however, we had to realize that existing ontologies are only partially suitable to model our data. Individual properties had definitions that did not match our data sets, so that no single ontology was found acceptable. Instead, we had to meticulously determine a set of ontologies whose parts would together cover most of our data. For the remaining portions we defined our own properties, with the intent to register the resulting ontology in the future.

The data modeling for the representation of researcher and publication utilizes several existing ontologies like the FOAF vocabulary and the Relationship Vocabulary. For subject headings the data modeling is based on the use of the Simple Knowledge Organization System (SKOS) and Dublin Core elements. We use 34 established properties and defined 250 properties ourselves. Table 4 lists the established ontologies we used.

Table 4. The established ontologies we used

Ontology	Namespace
dbo	http://dbpedia.org/ontology/
dcterms	http://purl.org/dc/terms/
foaf	http://xmlns.com/foaf/0.1/
iscover	http://i-scover.ieice.org/terms/iscover#
prism	http://prismstandard.org/namespaces/basic/2.0/
schema	http://schema.org/
skos	http://www.w3.org/2004/02/skos/core#
swrc	http://swrc.ontoware.org/ontology#
vcard	http://www.w3.org/2006/vcard/ns#

4.5 Data Enrichment

Data enrichment includes two steps, one is predicting missing values and the other one is link researcher, organization and keywords with DBpedia and Zhishi.me. For missing value prediction, we use a simple reasoning based method. More specifically, we use following rules to predict discipline and affiliation values. If the author of publication P is R, and author R's affiliation is A, then publication's affiliation is A. If the author of publication P is R, and author R's discipline is D, then publication's discipline is D. Table 5 shows the discipline and affiliation value complement results.

Table 5. The results of value complement in CAU library data

Data set	#Instance	#AddDiscipline	#AddAffiliation
SCI indexed journal paper	11934	9	875
Chinese journal paper	48449	12376	10927
Thesis	32755	567	5342
Patent	2389	1	936
Project	3941	756	2251
Monograph	1572	372	1238
Research results	536	2	498
Curriculum	410	8	52

<P author R> ∩ <R affiliation A> ⇒ <P affiliation A>

<P author R> ∩ <R discipline D> ⇒ <P affiliation D>

For researcher, organization and keyword linkage with DBpedia and Zhishi.me, we first conducts character and punctuations normalization, and then use normalized entity name as query to retrieval all the candidates from DBPedia and Zhishi.me. In order to obtain more accurate candidates, we conduct link analysis for each candidate. Specifically, if a candidate A has a redirect entity B, we add entity B into candidate set. If a candidate A is ambiguous, we add all the entities that candidate A may refer to into candidate set. After that, we use a ranking model that combines lexical and semantic similarity to determine which candidate should be linked. Specifically, we computes the string similarity between entity and each candidate using Levenshtein and Jaccard similarity. Semantic similarity is computed using semantic profiles. For organization, we use type and location information. For researcher, we use type, affiliation and research field. For keyword linkage, we use related keywords.

4.6 Republishing as Linked Data

The resources and properties in the library data namespace are published according to the Linked Data principles. The ontology contains all library data properties and class descriptions. Each resource is assigned a dereferenceable URI. The CAU linked library data includes 106109 resources in 10 classes, and 5826579 triples. We provide SPARQL endpoint at http://36.110.45.46:8890/sparql.

5   Conclusions and Future Work

In this paper we have presented an approach for integrating, enriching and republishing library data as Linked Data from several data sources including CAU library data, DBpedia and Zhishi.me. We have developed several components including a data cleaning, duplication detection and disambiguation, entity linkage and missing value prediction module. The linked library data includes 106109 resources in 10 classes, and 5826579 triples. A system with semantic search, statistic and visualization function is developed as well. We also conduct preliminary experiments and the results indicate the approach is effective.

Our future work include extensions of the presented data sets, methods, and the system itself. We plan to predict more missing values based on more sophisticated semantic reasoning methods. Cross-lingual data integration, e.g. linking English papers with researchers is another research direction.

References

1. Igata, N., Nishino, F., Kume, T., Matsutsuka, T.: Information integration and utilization technology using linked data. FUJITSU Sci. Tech. J. **50**(1), 3–8 (2014)
2. Krafft, D.B.: Linked data for libraries: a project update. In: 14th International Semantic Web Conference, United States of America, Bethlehem, pp. 11–15 (2015)
3. Monz, C., Weerkamp, W.: A comparison of retrieval-based hierarchical clustering approaches to person name disambiguation. In: 32nd International ACM SIGIR Conference on Research and Development in Information Retrieval, pp. 65–651 (2009)
4. Yoshida, M., Ikeda, M., Ono, S., Sato, I., Nakagawa, H.: Person name disambiguation by bootstrapping. In: 33th International ACM SIGIR Conference on Research and Development in Information Retrieval, pp. 10–17 (2010)
5. Xu, J., Lu, Q., Liu, Z.: Combining classification with clustering for web person disambiguation. In: 21st International Conference on World Wide Web, pp. 637–638 (2012)
6. Cucerzan, S.: Large-scale named entity disambiguation based on wikipedia data. In: 2007 Joint Conference on Empirical Methods in Natural Language Processing and Computational Natural Language Learning, pp. 708–716 (2007)
7. Damljanovic, D., Bontcheva, K.: Named entity disambiguation using linked data, In: 9th Extended Semantic Web Conference (2012)
8. Usbeck, R., Ngonga Ngomo, A.-C., Röder, M., Gerber, D., Coelho, S.A., Auer, S., Both, A.: AGDISTIS - graph-based disambiguation of named entities using linked data. In: Mika, P., Tudorache, T., Bernstein, A., Welty, C., Knoblock, C., Vrandečić, D., Groth, P., Noy, N., Janowicz, K., Goble, C. (eds.) ISWC 2014. LNCS, vol. 8796, pp. 457–471. Springer, Heidelberg (2014). doi:10.1007/978-3-319-11964-9_29
9. Mihalcea, R., Csomai, A.: Wikify! linking documents to encyclopedic knowledge. In: 17th ACM Conference on Information and Knowledge Management, pp. 233–242 (2007)
10. Milne, D., Witten, I.H.: Learning to link with wikipedia. In: 17th ACM Conference on Information and Knowledge Management, pp. 509–518 (2008)
11. Han, X.P., Sun, L.: A generative entity-mention model for linking entities with knowledge base. In: 49th Annual Meeting of the Association for Computational Linguistics: Human Language Technologies, vol. 1, pp. 945–954 (2011)
12. Ferragina, P., Scaiella, U.: TAGME: on-the-fly annotation of short text fragments. In: 19th ACM International Conference on Information and Knowledge Management, pp. 1625–1628 (2010)
13. Mendes, P.N., Jakob, M., García-Silva, A., Bizer, C.: DBpedia spotlight: shedding light on the web of documents. In: 7th International Conference on Semantic Systems, pp. 1–8 (2011)
14. Ratinov, L., Roth, D.: Design challenges and misconceptions in named entity recognition. In: 13th Conference on Computational Natural Language Learning, pp. 147–155 (2009)
15. Yosef, M.A., Hoffart, J., Bordino, I., Spaniol, M., Weikum, G.: AIDA: an online tool for accurate disambiguation of named entities in text and tables. In: PVLDB 2011, pp. 1450–1453 (2011)
16. Hu, W., Qu, Y., Cheng, G.: Matching large ontologies: a divide-and-conquer approach. Data Knowl. Eng. **67**(1), 140–160 (2008)
17. Jimenez-Ruiz, E., Grau, B.C., Zhou, Y.: Logmap 2.0: towards logic-based, scalable and interactive ontology matching. In: Ontology Matching, pp. 45–46 (2011)
18. Li, Y., Li, J.Z., Zhang, D., Tang, J.: Result of ontology alignment with RiMOM at OAEI 2006. In: Ontology Matching (2006)

19. Suchanek, F.M., Abiteboul, S., Senellart, P.: PARIS: probabilistic alignment of relations, instances, and schema. PVLDB 5(3), 157–168 (2011)
20. Bizer, C., Lehmann, J., Kobilarov, G., Auer, S., Becker, C., Cyganiak, R., Hellmann, S.: DBpedia - a crystallization point for the web of data. J. Web Semant. 7, 154–165 (2009)
21. Niu, X., Sun, X., Wang, H., Rong, S., Qi, G., Yu, Y.: Zhishi.me - weaving chinese linking open data. In: Proceedings of 10th International Semantic Web Conference, Bonn, pp. 23–27 (2011)

Object Clustering in Linked Data Using Centrality

Xiang Zhang[1(✉)], Yulian Lv[2], and Erjing Lin[1]

[1] School of Computer Science and Engineering,
Southeast University, Nanjing, China
{x.zhang,linerjing}@seu.edu.cn
[2] College of Software Engineering (Suzhou),
Southeast University, Suzhou, China
lvyulian@seu.edu.cn

Abstract. Large-scale linked data is becoming a challenge to many Semantic Web tasks. While clustering of graphs has been deeply researched in network science and machine learning, not many researches are carried on clustering in linked data. To identify meta-structures in large-scale linked data, the scalability of clustering should be considered. In this paper, we propose a scalable approach of centrality-based clustering, which works on a model of Object Graph derived from RDF graph. Centrality of objects is calculated as indicators for clustering. Both relational and linguistic closeness between objects are considered in clustering to produce coherent clusters.

1 Introduction

The great volume of linked data is becoming a challenge for many Semantic Web tasks. These tasks vary from semantic query [1] to semantic mining [2]. The scale of linked data demands new methods to discover knowledge from the links or linguistics in linked data. A promising approach is to decompose linked data into clusters, which are sets of densely inter-connected objects. The identification of these clusters is of crucial importance as they may help to scale down the problem when exploring linked data, or may help researchers to understand the meta-structure of the linked data.

Clustering approaches have been deeply researched in the modern science of networks and machine learning. While clustering approaches like K-means or spectral clustering are commonly used and effective in small or medium dataset, they can be hardly adapted to the scale of linked data. To the best of our knowledge, clustering or community detection in linked data is still a research area not being deeply explored. There are two major problems facing this area: (1) A near-linear clustering approach is needed to efficiently decompose massive linked data; (2) How to effectively utilize relations and linguistic information of objects, which are both abundant in linked data.

We propose a centrality-based clustering in this paper, which is efficient for clustering large-scale linked data. We introduce Object Graph as the graph model. The closeness between two objects is measured both relationally and linguistically. The notion of Virtual Document is used to measure linguistic closeness between objects. For each object in linked data, a set of graph centralities is assessed and k centroids are

H. Chen et al. (Eds.): CCKS 2016, CCIS 650, pp. 172–183, 2016.
DOI: 10.1007/978-981-10-3168-7_17

selected using a distance-maximization strategy. An LPA-based clustering will decompose linked data into k clusters.

2 Models and Architecture

In this section, we propose Object Graph as the graph model for clustering. A Virtual Document is built for each object in Object Graph to capture its linguistic information. The architecture of our approach is also discussed.

2.1 Object Graph and Virtual Document

RDF model of linked data is multi-mode and multi-dimensional with multiple types of nodes (classes, properties, objects or literals) and multiple types of relations. It is not suitable for object clustering. We propose a single-mode and single-dimensional graph model, called Object Graph, as the graph model for object clustering.

Definition 1 (Object Graph): Given a Linked Data ℓ, its Object Graph $\mathcal{G}(\ell) = \langle \mathcal{O}, \mathcal{W}, \mathcal{V}d_n \rangle$ is a directed graph. $\mathcal{O}$ is the node set, which comprises all the named objects defined or referred in ℓ; $\mathcal{W}$ is a weighting scheme of edges. Given $o_i, o_j \in \mathcal{O}$, if $\mathcal{W}(o_i, o_j) > 0$, there is a weighted edge from o_i to o_j in ℓ. $\mathcal{W}(o_i, o_j)$ equals to the closeness from o_i to o_j. $\mathcal{V}d_n$ is a labeling function of $\mathcal{G}(\ell)$. For each $o_i \in \mathcal{O}$, $\mathcal{V}d_n(o_i)$ is called n-step virtual document of o_i, which is a bag of words capturing linguistic information of o_i in ℓ.

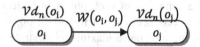

Fig. 1. The model of object graph

Shown in Fig. 1, each node in Object Graph represents a named object, and there is an edge from one object to another when (1) there is a direct relation between them in RDF model; (2) or there is a directed path between them, and all intermediate objects are blank nodes. Thus, Object Graph captures all direct relations between named objects, and also captures indirect relations formed by blank nodes. The edges are weighted by closeness between objects.

Definition 2 (Object Description): Given an object o_i in linked data ℓ, the object description of o_i in ℓ is a bag of words defined by Eq. (1):

$$d(o_i) = \cup \{d_{\text{uri}}(o_i), d_{\text{labl}}(o_i), d_{\text{comm}}(o_i), d_{\text{anno}}(o_i)\} \tag{1}$$

In Eq. (1), $d_{\text{uri}}(o_i)$ contains words in the URI of o_i; $d_{\text{labl}}(o_i)$ and $d_{\text{comm}}(o_i)$ are words occurred in *rdfs:label* and *rdfs:comment* properties of o_i respectively; $d_{\text{anno}}(o_i)$ is the words from other annotation properties of o_i. $\cup$ is the operation of merging bags of words.

Definition 3 (Virtual Document): A virtual document $Vd_n(o_i)$ is a bag of words encapsulating the linguistic information of object o_i and its n-step surrounding neighbors. The 0-step Virtual Document of o_i $Vd_0(o_i) = d(o_i)$.

$$neighbor_n(o_i) = \overleftarrow{neighbor_n}(o_i) \cup \overrightarrow{neighbor_n}(o_i) \tag{2}$$

$$Vd_n(o_i) = \cup_{o_j \in neighbor_n(o_i)} d(o_j) \tag{3}$$

In Eqs. (2 and 3), $\overrightarrow{neighbor_n}(o_i)$ and $\overleftarrow{neighbor_n}(o_i)$ represent the set of objects that o_i can access through a forward or backward n-step links. $Vd_n(o_i)$ is the virtual document of o_i comprising all object descriptions of itself and its n-step neighbors.

The notion of virtual document is originated from [3], which aimed at capturing linguistic information for ontology matching. While an object description provides firsthand but limited information about the semantics of an object, a virtual document is a comprehensive and abundant corpus to characterize the object.

2.2 Architecture

As shown in Fig. 2, our approach of clustering is architected into three layers. The Modeling Layer uses an RDF parser to get the RDF model of a linked data as input. Then virtual document of each object is then extracted, and the Object Graph is constructed from RDF model. Derived Object Graph will be passed to Analysis Layer, whose major task is to calculate the relational and linguistic closeness between objects, or in other words, to refine the edge weights of Object Graph. The last layer, Clustering Layer, will first assess the centrality of each object in Object Graph, then utilize the centrality as an indicator to produce a set of important object as centroid candidates. k centroids are selected using a distance-maximization strategy. For each centroid, an LPA-based clustering will be carried to produce clusters. Finally, isolated objects and sub-graphs will be merged into k clusters.

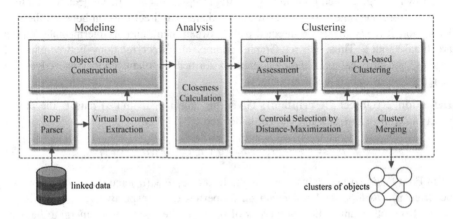

Fig. 2. Architecture of centrality-based clustering

3   Closeness Calculation

In linked data, two objects are deemed to be close in two ways: (1) They are close if
there is an explicit statement that they have a relation. For example: a student who
knows another student. (2) They are similar in semantics, which can be captured in
their linguistic information, even if they don't have a direct relation. For example, two
researchers can be semantically close when there is no co-authorship, but the textual
descriptions of them indicate that they are quite similar in research interests.

In addition to relations, linguistic similarity in linked data is an important indicator
for clustering of objects. Some Semantic Web tasks rely on the analysis of object
descriptions, such as entity linking from unstructured text to semantic objects. These
tasks will benefit if linguistically close objects can be grouped together. Besides,
objects with similar descriptions are possible to develop a potential relation in the
future, such as the two researchers with same research interests. In our approach,
linguistic closeness will affect the clustering in three aspects: the weighting of edges in
Object Graph, the LPA-based clustering of objects and the merge of isolated objects
and sub-graphs into clusters.

The relational part of closeness $\mathcal{W}_r(o_i, o_j)$ is calculated by Rule 1 and 2. The
linguistic part of closeness $\mathcal{W}_l(o_i, o_j)$ is calculated by Eq. (4). Finally, edge weights in
Object Graph is calculated as the multiply of the two parts as shown in Eq. (5). Given
linked data ℓ:

Rule 1: For each $o_j \in \overrightarrow{neighbor}_1(o_i)$ or $o_k \in \overleftarrow{neighbor}_1(o_i)$ in ℓ, there is a directed
edge from o_i to o_j or from o_k to o_i in $\mathcal{G}(\ell)$. $\mathcal{W}_r(o_i, o_j)$ or $\mathcal{W}_r(o_k, o_i)$ equals to the
number of distinct relations from o_i to o_j or from o_k to o_i respectively.

Rule 2: For each $o_j \in \overrightarrow{neighbor}_n(o_i)$ or $o_k \in \overleftarrow{neighbor}_n(o_i)$ in ℓ, if all intermediate
nodes lie on the n-step path from o_i to o_j or from o_k to o_i are blank nodes, $\mathcal{W}_r(o_i, o_j) =
1/n$ or $\mathcal{W}_r(o_k, o_i) = 1/n$ respectively.

$$\mathcal{W}_l(o_i, o_j) = cos\theta = \frac{\overrightarrow{Vd_n(o_i)} \cdot \overrightarrow{Vd_n(o_j)}}{\|Vd_n(o_i)\| \cdot \|Vd_n(o_j)\|} \tag{4}$$

$$\mathcal{W}(o_i, o_j) = \mathcal{W}_r(o_i, o_j) \times \mathcal{W}_l(o_i, o_j) \tag{5}$$

In Eq. (4), $\overrightarrow{Vd_n(o_i)}$ is the term vector of n-step virtual document of o_i, and $\|Vd_n(o_i)\|$
is the document length.

4   Centrality Assessment

The centrality measurements are to find the potential of objects to be centroids of
clusters. Heuristically, objects with high centrality are more likely and adequate to be
the center of a cluster, comparing to ones with low centrality.

Various notions of centrality and their measurements have been proposed in literals. They can be classified into three categories: Degree centrality, Shortest-Path-based centrality and Eigenvector centrality.

Degree is a simple yet powerful measurement of objects' centrality in Object Graph. Relations between objects can be seen as conferral of importance. Objects with high degree centrality are intuitively important in the graph since they receive many conferral of importance from others. In our approach, degree centrality of object o_i is noted as $C_D(i)$.

Shortest-Path-based centrality is a set of notions based on shortest paths linking pairs of vertices, such as the Betweenness Centrality [4] measured by the ratio of shortest paths across it in Object Graph. The calculation of Shortest-Path-based centralities usually has a high computational complexity, which makes it difficult to adapt to big data, such as linked data. Besides, this category of centralities doesn't outperform degree centrality in some Semantic Web tasks, such as stated in [5]. Considering the scalability, Shortest-Path-based centrality is not adopted in our approach.

The calculation of eigenvector centrality is based on finding the eigenvector of the adjacency matrix encoding a graph. Two well-known measurements of eigenvector centrality on the Web are PageRank [6] and HITS [7]. PageRank is used by the Google search engine for ranking web pages. The authority of a page is computed recursively as a function of the authorities of the pages that link to it. HITS computes two values related to topological properties of the Web pages, the "authority" and the "hubness". In our approach of clustering, three weighted variations of PageRank and HITS are used to define the eigenvector centrality of objects in linked data.

In Eqs. (6–9), $C_{PR}(i)$ is the original PageRank centrality. $C_{WPR}(i)$ is an weighted extension to $C_{PR}(i)$. $C_{HITS-A}(i)$ and $C_{HITS-H}(i)$ are the weighed extension of the authority and hubness in HITS algorithm. In the calculation of weighted HITS, the symbol $\|x\|$ means the normalization of x after each iteration.

$$C_{PR}(i) = \frac{1-d}{|O|} + d \times \sum_{j \in \overrightarrow{neighbor_1}(i)} \frac{C_{PR}(j)}{|\overrightarrow{neighbor_1}(j)|} \qquad (6)$$

$$C_{WPR}(i) = \frac{1-d}{|O|} + d \times \sum_{j \in \overrightarrow{neighbor_1}(i)} \frac{w(o_j,o_i) \times c_p(j)}{\sum_{j \in \overrightarrow{neighbor_1}(i)} w(o_j,o_i)} \qquad (7)$$

$$C_{HITS-A}(i) = \left\| \sum_{j \in \overrightarrow{neighbor_1}(i)} w(o_i,o_j) \times C_{HITS-H}(j) \right\| \qquad (8)$$

$$C_{HITS-H}(i) = \left\| \sum_{j \in \overrightarrow{neighbor_1}(i)} w(o_j,o_i) \times C_{HITS-A}(j) \right\| \qquad (9)$$

5   Centroid Selection and Clustering

Centrality of objects indicates their topological and topical importance in linked data. An object with high centrality is usually a center object surrounded by a set of close-neighboring objects. With a set of selected centroids, the huge amount of objects in a given linked data can be clustered based on the distance between centroids and non-centroids, which is the basic idea of many clustering algorithms, such as the commonly used K-means clustering.

A naïve strategy to find centroids is to simply select top-ranked objects according to their centralities. Given k as an expected cluster numbers, top-k objects with high centralities will be chosen as centroids. However, there is a well-known TKC (Tightly-Knit Community) effect stated in [8], which could make the centrality-based clustering problematic. Objects in a tightly-knit community will mutually reinforce their centralities and dominate the set of top-k selected centroids. A clustering based on these centroids will result in a poor coverage on the whole dataset. In our approach, a set of 10 k of candidate centroids will be selected beforehand according to their centrality. This enlarged candidate set contains all possible centroids to be further selected. A distance-maximization strategy is proposed in Algorithm 1, in which k centroids are selected one by one considering their distance to pre-selected centroids. The goal of this strategy is to maximize the mutual distance among centroids in linked data, to fulfill a well-covered clustering of objects.

Algorithm 1 : Distance-maximization Strategy for Centroid Selection

---

Input: a set of objects $\mathcal{O}$ with centrality values $\mathcal{C}$, parameter k as the expected number of clusters.
1. Set the set of centroid $\mathcal{O}_c$ to an empty set;
2. Rank the set of objects $\mathcal{O}$ in descending order according to $\mathcal{C}$;
3. Select top $10k$ objects in $\mathcal{O}$ to form a set of centroid candidates: $\mathcal{O}_{c\prime}$;
4. $\mathcal{O}_c \leftarrow \{ o_i \mid o_i \in \mathcal{O}_{c\prime}$, and o_i has top centrality in $\mathcal{O}_{c\prime}\}$;
5. $\mathcal{O}_{c\prime} \leftarrow \mathcal{O}_{c\prime}/\{o_i\}$;
6. Repeat, until $|\mathcal{O}_c| = k$:
    a)   Find o_i in $\mathcal{O}_{c\prime}$, $o_i = argmax \sum_{o_j \in \mathcal{O}_c} d(o_i, o_j)$
    b) $\mathcal{O}_c \leftarrow \mathcal{O}_c \cup \{o_i\}$;
    c) $\mathcal{O}_{c\prime} \leftarrow \mathcal{O}_{c\prime}/\{o_i\}$;

Output: the set of centroids $\mathcal{O}_c$

In Algorithm 1, $d(o_i, o_j)$ represents the distance between o_i and o_j. Its calculation is shown in Eq. (9), in which $o_k, o_l \in \rho(o_i, o_j)$ means o_k, o_l lie on a shortest path $\rho(o_i, o_j)$ between o_i and o_j:

$$d(o_i, o_j) = \sum_{o_k, o_l \in \rho(o_i, o_j)} 1 - \mathcal{W}(o_k, o_l) \tag{10}$$

After centroid selection, all non-centroids will be grouped into k clusters. An LPA-based (Label Propagation Algorithm) clustering is proposed in Algorithm 2. Each centroid will propagate its cluster label to neighboring objects iteratively until no more objects can be reached. Different with the original LPA, when a non-centroid object is propagated with multiple labels during the iteration, its label will be judged to the cluster whose centroid has the greatest linguistic closeness to it.

Algorithm 2 : LPA-based Clustering

Input: the set of centroids $\mathcal{O}_c = \{o_1, o_2, ..., o_k\}$
1. Initially set $\mathcal{O}_1 \leftarrow \{o_1\}$, $\mathcal{O}_2 \leftarrow \{o_2\}$, ..., $\mathcal{O}_k \leftarrow \{o_k\}$
2. Repeat, until no more object can be merged into $\mathcal{O}_1$ to $\mathcal{O}_k$:
    a)   For each $\mathcal{O}_p \in \{o_1, ..., o_k\}$,
        i.   For each $o_i \in \mathcal{O}_p$, find $\mathcal{O}_p' = \mathcal{O}_1 \cup \overrightarrow{neighbor_1}(o_i) \cup \overleftarrow{neighbor_1}(o_i)$;
        ii. For each $o_j \in \mathcal{O}_p'$, label o_j with a cluster id: p
    b)   For each non-centroid object o_j, and o_j has been labeled with multiple cluster ids, re-label its cluster id with the cluster whose centroid has the greatest $\mathcal{W}_l$ to o_j.
    c)   For those labeled non-centroid objects, merge them into corresponding clusters according to their cluster ids.
3. For each remaining non-centroid objects (isolated objects or sub-graphs, etc.), classify them into k clusters according to $\mathcal{W}_l$.

Output: A clustering of $\mathcal{O}$ into k clusters: $\mathcal{O}_1, \mathcal{O}_2, ..., \mathcal{O}_k$

Considering there may be isolated objects or sub-graphs remained after clustering, the step 3 of Algorithm 2 will finally merge them into the k clusters. The merging of remaining objects is basically a text classification problem, which utilize the linguistic closeness between each remaining object and k centroids. We omit the details of merging for the sake of conciseness.

6 Evaluation

In this section, we first analysis the datasets, then evaluate the performance of different centrality measurements and the final clustering. We carried out these evaluations on our server with Intel Xeon E3 V2 processors and 16G RAM.

6.1 Datasets

Three linked data are selected as the dataset of experiments, i.e., (1) Semantic Web Conference Corpus (SWCC in short)[1], which is a data on Semantic Web Conference; (2) Jamendo (JAME in short)[2], which is a data on licensed music; (3) LinkedMDB (LMDB in short)[3], which is a data for movies; In Table 1, the statistics of each dataset is presented. #triple is the total number of triples; #object is the number of objects;

---

[1] SWCC: http://data.semanticweb.org/.

[2] JAME: http://dbtune.org/jamendo/.

[3] LMDB: http://linkedmdb.org/.

Table 1. Statistics of each linked data

Data	#triple	#object	#class	#property	#relation
SWCC	20,802	3,089	20	77	10,589
JAME	1,049,647	412,565	12	26	505,961
LMDB	6,247,909	1,326,885	53	222	2,069,454

#class and #properties represent the number of classes and properties that the dataset used as vocabulary; #relation is the number of object links, which is also the number of edges in Object Graph.

Figure 3 shows the abundance of linguistic information in each dataset. In Fig. 3(a) and (b), the X axis respectively represents the number of unique words in a certain object's 1-step virtual document, and the document length of virtual document. In both figures, the Y axis represents the percentage of objects whose linguistic information is equal to or more abundant than a given value. A median line is drawn to illustrate the average linguistic abundance in each dataset. From both figures we can observe that the SWCC has the most abundant linguistic information, while the JAME has the least.

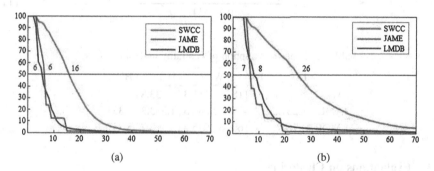

Fig. 3. Statistics of linguistic abundance on (a) unique word (b) virtual document length

6.2   Evaluations on Centrality Assessment

To evaluate which measurement will produce the most reasonable candidate set, a prior ground true of human judgment should be generated, and the agreement among human-generated and machine-generated candidate sets will be calculated to find the best measurement, as stated in [5]. However, for the evaluation on large-scale linked data, the generation of ground true by human is impossible. Instead, we use the agreement among five machine-generated centralities, as well as their time performances, as selectors to filter out three measurements for the final clustering.

We use Kendall's tau statistic [9] to calculate the correlation among ranked candidate sets produced by degree centrality (*DE* in short), PageRank centrality (*PR*), Weighted PageRank (*WPR*), HITS-authority (*HA*) and HITS-hubness (*HH*). The calculation is shown in Eq. (11), where the correlation τ is the odds that two objects are ranked concordantly against discordantly in two candidate sets. The agreements among five centralities are shown in Table 2.

Table 2. Agreement between various centralities

SWCC	DE	PR	WPR	HA	HH	JAME	DE	PR	WPR	HA	HH	LMDB	DE	PR	WPR	HA	HH
DE	1	-0.08	-0.21	-0.10	-0.23	DE	1	-0.21	-0.31		-0.19	DE	1	-0.11	-0.05	0.35	-0.05
PR	-0.08	1	-0.16	-0.23	-0.29	PR	-0.21	1	0.13	-0.21	-0.15	PR	-0.11	1	0.19	-0.03	-0.25
WPR	-0.21	-0.16	1	-0.38	-0.44	WPR	-0.31	0.13	1	-0.31	-0.18	WPR	-0.05	0.19	1	-0.02	-0.25
HA	-0.10	-0.23	-0.38	1	-0.06	HA	1	-0.21	-0.31	1	-0.11	HA	0.35	-0.03	-0.02	1	-0.10
HH	-0.23	-0.29	-0.44	-0.06	1	HH	-0.19	-0.15	-0.18	-0.11	1	HH	-0.05	-0.25	-0.25	-0.10	1

We use Gephi 0.9.1[4] as our tool for centrality assessment. The time performance of each measurement is shown in Table 3. From the results in both tables, we select *DE*, *WPR* and *HA* as the final measurements to produce centroid candidates. *DE* is selected because its simpleness and efficiency in calculation. *WPR* is selected because it concerns linguistic information in centrality assessment and shows a difference with non-weighted PageRank. *HA* is selected because it shows a good correlation with *DE* on two datasets, and also has a sound time performance.

$$\tau = \frac{\#concordant - \#discordant}{n(n-1)/2} \tag{11}$$

Table 3. Time consumption of centrality assessment (ms)

	DE	PR	WPR	HA	HH
SWCC	36.8	118	111.3	33.8	33.8
JAME	1,237.5	2,435	3,086.5	1,333.3	1,333.3
LMDB	4,765	24,160	39,045.2	5,557.7	5,557.7

6.3   Evaluations on Clustering

After the generation of centroid candidates, k centroids will be selected and the dataset will be decomposed into k clusters. To evaluate the performance of clustering, we use K-means as the baseline clustering algorithm. Weka 3 is used as our tool for K-means clustering. We use *Connectedness* defined in [10] as the indicator for the quality of clustering, which is commonly used in the evaluation of ontology modularization. The calculation of *Connectedness* is shown in Eq. (12), where $|E_x|$ is the number of shared edges in between clusters, and $|E|$ is the number of all edges.

$$connectedness = \frac{|E_x|}{|E|} \tag{12}$$

Table 4 shows the resulted quality of clustering. Both *DE*, *WPR* and *HA* produce high-quality clusters with our LPA-based clustering algorithm. The average performance on all datasets indicates that *WPR* is the best choice comparing to other two

---

[4] Gephi: https://gephi.org/users/download/.

Table 4. Quality evalution of different clusterings

	K-means	DE	WPR	HA
SWCC	0.203	0.122	0.120	0.113
JAME	–	0.021	0.023	0.021
LMDB	–	0.014	0.010	0.056
Avg.	–	0.052	0.048	0.063

measurements, and it produces clustering with less than 5 percents of shared edges in between clusters. As we expected, K-means failed to decompose JAME and LMDB because of its computational complexity and the data volume. K-means only successfully decomposed SWCC with a *connectedness* of 0.203, which indicates a much lower quality of clustering comparing to our approach.

7 Related Works

To the best of our knowledge, clustering or community detection in linked data is still a research area not being deeply explored. Grimnes et al. presented in [11] several ways to extract instances from RDF graph and computing the distance between them. The challenge surrounding the application of clustering algorithms to Semantic Web data was also discussed. Yan proposed RDF graph partitioning in [12], in which large RDF graph would be partitioned into sub-graphs and stored individually. In [13], Aluc proposed RDF clustering for RDF data management. They kept track of RDF records in DB that are co-accessed by queries in the workload and physically clustered them. These works differs with our approach that their goal of portioning RDF graph is to fulfill a self-adaptive RDF management to improve the efficiency of SPARQL query, while our approach aims at discovering meta-structure of linked data for diverse Semantic Web tasks.

Although object clustering hasn't been fully discussed in Semantic Web research community, centrality-based clustering on large-scale graphs has been discussed in the research of network science. Tabrizi proposed in [14] a personalized PageRank clustering based on random walks, which has a linear time and space complexity. The basic idea of this work is similar to ours. Since the dataset of this work is web pages, our work differs with it in many aspects: the graph model, the calculation of closeness, the centroid selection strategy and the clustering algorithm. However, it motivates us and proves that centrality-based clustering in large-scale linked data is feasible.

8 Conclusion and Future Work

The identification of object clusters in linked data is of crucial importance as they may help to scale down the problem when exploring linked data, or may help researchers to understand the meta-structure of the linked data. We propose an efficient centrality-based object clustering in this paper. Object Graph is introduced as the graph model of clustering. The closeness between two objects is measured in both relational and linguistic manner. A distance-maximization strategy is used to select centroids from candidates with high centrality. An LPA-based clustering decomposes linked data into k clusters. Our experiments show that our approach is feasible in large-scale linked data.

In our future work, we will explore the possibility of a guided clustering, in which object clustering will be guided by ontology modularization. The modules in TBox may provide information about how different types of objects are related. We will also try to performance our clustering on larger linked data, such as DBpedia. A visualized system of object clusters will be constructed for better human understanding.

Acknowledgement. The work was supported by the National High-Tech Research and Development (863) Program of China (No. 2015AA015406) and the Open Project of Jiangsu Key Laboratory of Data Engineering and Knowledge Service (No. DEKS2014KT002).

References

1. Hartig, O., Bizer, C., Freytag, J.-C.: Executing SPARQL queries over the web of linked data. In: Bernstein, A., Karger, D.R., Heath, T., Feigenbaum, L., Maynard, D., Motta, E., Thirunarayan, K. (eds.) ISWC 2009. LNCS, vol. 5823, pp. 293–309. Springer, Heidelberg (2009). doi:10.1007/978-3-642-04930-9_19
2. Paulheim, H.: Exploiting linked open data as background knowledge in data mining. In: Proceedings of International Workshop on Data Mining on Linked Data, with Linked Data Mining Challenge Collocated with ECMLPKDD 2013, pp. 1–10 (2013)
3. Qu, Y., Hu, W., Cheng, G.: Constructing virtual documents for ontology matching. In: Proceedings of 15th International Conference on World Wide Web (WWW 2006), pp. 23–31 (2006)
4. Newman, M.E.J.: A measure of betweenness centrality based on random walks. Soc. Netw. **27**, 39–54 (2005)
5. Zhang, X., Cheng, G., Qu, Y.: Ontology summarization based on RDF sentence graph. In: Proceedings of 16th International Conference on World Wide Web – WWW 2007, p. 707 (2007)
6. Page, L., Brin, S., Motwani, R., Winograd, T.: The PageRank citation ranking: bringing order to the web. Technical report, Stanford Digital Library Technologies Project (1998)
7. Kleinberg, J.M.: Authoritative sources in a hyperlinked environment. J. ACM **46**, 668–677 (1999)
8. Lempel, R., Moran, S.: Stochastic approach for link-structure analysis (SALSA) and the TKC effect. Comput. Netw. **33**, 387–401 (2000)
9. Sheskin, D.J.: Handbook of parametric and nonparametric statistical procedures. Technometrics **46**, 1193 (2004)
10. Schlicht, A., Stuckenschmidt, H.: Towards structural criteria for ontology modularization. In: CEUR Workshop Proceedings (2006)
11. Grimnes, G.A.A, Edwards, P., Preece, A.: Instance based clustering of semantic web resources. In: Proceedings of 5th European Semantic Web Conference on the Semantic Web: Research and Applications, pp. 303–317 (2008)

12. Yan, Y., Wang, C., Zhou, A., Qian, W., Ma, L., Pan, Y.: Efficient indices using graph partitioning in RDF triple stores. In: Proceedings - International Conference on Data Engineering, pp. 1263–1266 (2009)
13. Aluç, G., Özsu, M.T., Daudjee, K.: Clustering RDF databases using tunable-LSH. pp. 1–13, CoRR, abs/1504.02523 (2015)
14. Tabrizi, S.A., Shakery, A., Asadpour, M., Abbasi, M., Tavallaie, M.A.: Personalized PageRank clustering: a graph clustering algorithm based on random walks. Phys. A Stat. Mech. Appl. **392**, 5772–5785 (2013)

Research on Knowledge Fusion Connotation and Process Model

Hao Fan[1], Fei Wang[1(✉)], and Mao Zheng[2]

[1] School of Information Management,
Wuhan University, Wuhan 430072, Hubei, People's Republic of China
feiwang@whu.edu.cn
[2] Department of Computer Science,
University of Wisconsin-La Crosse, La Crosse, WI 54601, USA

Abstract. The emergence of big-data brings diversified structures and constant growths of knowledge. The objective of knowledge fusion (KF) research is to integrate, discover and exploit valuable knowledge from distributed, heterogeneous and autonomous knowledge sources, which is the necessary prerequisite and effective approach to implement knowledge services. In order to apply KF practice, this paper firstly discusses KF connotations in terms of analysing the relations and differences among various notions, i.e. knowledge fusion, knowledge integration, information fusion and data fusion. Then, based on the knowledge representation method using ontology, this paper investigates several KF implementation patterns and provides two types of dimensional KF process models oriented to demands of knowledge services.

Keywords: Knowledge fusion · Knowledge representation · Fusion pattern · Process mode

1 Introduction

With the development of data creating, releasing, storing and processing technologies, data is showing a rapid growth trend in all society areas. Of all the data available to the human civilization, 90% were produced in the past two years, the big data era has arrived [16]. Knowledge is awareness and understanding about people or things in the objective world, which is generated by feeling, communicating and logic inference activities in the course of practice and education and maybe facts, information or skills. The information chain, formed with "fact → data → information → knowledge → wisdom", indicates that big data contains huge amount of information, from which large knowledge can be extracted. Big data gives rise to the emergence of large scale knowledge bases. Famous knowledge base research projects, e.g. DBpedia, KnowItAll, NELL and YAGO, use information extraction techniques acquiring knowledge from high quality network data sources (e.g. Wikipedia), and automatically realize its construction and management [22]. Meanwhile, big data brings about information overload

© Springer Nature Singapore Pte Ltd. 2016
H. Chen et al. (Eds.): CCKS 2016, CCIS 650, pp. 184–195, 2016.
DOI: 10.1007/978-981-10-3168-7_18

and pollution too, in which knowledge presents characteristics of heterogeneity, diversity and independence. In the era of data, with rapidly increasing of information and knowledge, knowledge discovery has become the research focus in various disciplines, including data science and information science [25]. Therefore, in order to improve the efficiency and quality of knowledge service, issues of analysing and utilizing knowledge existing in big data, eliminating the inconsistency between different knowledge sources, and extracting, discovering and inducing the potential valuable connotations, have become important in knowledge management research.

The continuous formation and evolution have brought about autonomous, heterogeneous and multi-source features of knowledge. Knowledge Fusion (KF) is a process of acquiring and utilizing knowledge aiming at the problem of knowledge service. Operated by KF activities, implicate and undiscovered valuable knowledge is mined from various distributed and heterogeneous data sources. KF converts autonomous knowledge into new one with higher levels of intension and reliability, helps users to find potential associations between knowledge and fact, and improves decision-making levels by making more efficient, objective and scientific judgments. KF becomes a new growth point for knowledge service [23].

As an important part of knowledge management and engineering, KF has been widely received the attention of scholars in many fields, such as computer science, knowledge engineering and information science. Smirnov et al. [21] investigates patterns for context-based KF In the decision support systems. Dong et al. [7] analyses differences and relations between data fusion and KF, and realizes KF processes by combining knowledge extraction and traditional data fusion methods together. Tang and Wei [23] discusses the requirement of big data KF and its basic framework. Liu et al. [15] defines a structure of multi-domain ontology and provides dynamic ontology based on KF demands through mappings between different domain ontologies. Xu et al. [24] designs a KF framework based on ontology, which is consists of several parts, such as constructing meta knowledge set, determining knowledge measurement indicators, designing fusion algorithm, applying fused knowledge, and so on. Qiu and Yu [20] summaries the KF implementation path as four types based on semantic rules, *Bayesian* networks, *D-S* theories and knowledge mining, with which Zhou et al. [26] discusses various KF processing algorithms. Guo et al. [9] reviews and evaluates research trends and theoretical developments of KF, and indicates that, there is not yet a formed general framework for KF systems, as well as directly applicable KF algorithms and standardized KF procedures. The existing research mainly focuses on specific KF frameworks, algorithms, and practical theories.

In terms of time distribution of related literatures, KF is a new research topic which is produced with the change of knowledge service requirements and the development of knowledge management research. In order to implement KF in practice, it is necessary to correctly understand KF connotation by analysing relations and differences among various relative notions, i.e. knowledge fusion, knowledge integration, information fusion and data fusion, and analyse KF implementation patterns and its process models.

2   Knowledge Fusion Connotation

2.1   Conception of Knowledge Fusion

KF is a new concept developed on the basis of information fusion. There are many intersections between the two research areas. The early definition of KF is given by Preece in the KRAFT project [19], refers to a process locating and extracting knowledge from multiple, heterogeneous on-line sources and transforming it so that the union of the knowledge can be applied in problem-solving. The KF system in KRAFT project includes three layers of services: knowledge retrieval, transformation and fusion, in which KF is defined to associate, link and simplify the transformed distributed knowledge with a unified model, and provide solutions for the problem under specific conditions.

Smirnov et al. [21] proposes that the aim of KF is to integrate multi-source information and knowledge into a unified knowledge structure model, in order to allow decision-makers to understand and look insight into the decision-making environment and provide the needed knowledge to solve problems. Hou et al. [11] and Xu et al. [24] believe that KF is the process of intelligently processing distributed databases, knowledge bases and data warehouses, and acquiring new knowledge by transformation and integration procedures. It aims to realize the sharing and cooperation between different knowledge resource systems, and apply knowledge mining among knowledge bases. These definitions have carried on the inheritance and development to the Preece's KF concept, which is emphasized that fusion results are productions of new knowledge.

Guo et al. [9] and Tang and Wei [23] propose that KF is mainly studying the transformation, integration and aggregation processes in distributed knowledge base systems in order to generate new knowledge, and investigating optimization processes of knowledge structures and contents to provide knowledge service. This definition concerns processes of knowledge innovation and knowledge optimization, indicates the KF aim as providing knowledge services, and extends the KF object from traditional resources (such as databases, knowledge bases, fact parameters acquired by sensors, etc.) to the one including rules, models, methods, and even experiences, ideas, etc. In other words, the object of KF includes not only explicit knowledge, but also tacit knowledge.

Dong and Srivastava [8] considers KF as the issue assessing and measuring the accuracy of extracting knowledge. In the process of building a knowledge base, it is required to extract knowledge from distributed data sources, and integrate it into the base. A number of different knowledge extractors might be used during knowledge extraction, and each extractor generates its corresponding knowledge results. So, it is required to evaluate the accuracy of each extracted result to improve the correctness of knowledge bases.

Hu and Cao [10] extracts and transforms sentences in Web page texts into triple semantic nets for representing knowledge. It defines KF as the process eliminating contradictions among extracted knowledge and integrating its structures in accordance with user constraints and rules, which solves problems of incomplete, fuzzy, redundant and inconsistent knowledge contained in Web page texts.

Kampis and Lukowicz [12] proposes the notation of *Collaborative KF*, and indicates that traditional KF assumes informational completeness, while collaborative KF is a version of KF where traditional fusion events are local, e.g. happen upon the meetings of individual knowledge providers, and global fusion happens due to the collective (hence "collaborative") interaction dynamics. In collaborative KF, there is no guarantee that different knowledge sources were keeping unchanged and available at any time.

To sum up, concepts of KF are different in different periods and research fields. In the field of computer science and database research, KF emphasizes on the representation, transformation, cleansing and integration of explicit knowledge, focuses on eliminating the inconsistency, incompleteness, redundancy and uncertainty of knowledge among different knowledge sources, which mainly investigates on KF algorithm design and implementation so as to improve the standardization and credibility of fused knowledge. In the field of library and information science, knowledge refers to the sum of cognition and experience in the practice of changing the world, in which both explicit knowledge and tacit knowledge are concerned. KF research is to construct theory and method systems, which emphasizes on the integration of tacit knowledge and its impact.

2.2 Knowledge Fusion and Knowledge Integration

KF and knowledge integration are both knowledge object-oriented in terms of dealing with different structure and multi-source knowledge, which have connections and differences to each other. Literally, "integration" is the process of aggregating multiple individual objects to form a whole one, while "fusion" is the process of recombining multiple individual objects, splitting and dismantling it into a complete one. Integration emphasizes on aggregation and combination, while fusion more on merging and reorganizing. After fusion process, knowledge objects are supposed to have new emerging features relative to original ones.

Scholars have given definitions of knowledge integration from various perspectives. In the field of management, library and information science, Liu and An [13] indicates that knowledge integration refers to the process of dynamically enhancing the core competitiveness of an organization though different merging levels between knowledge and knowledge, knowledge and people, and knowledge and procedures, which aims to realize the knowledge innovation. Cai and Chen [6] gives a review of knowledge integration research, and proposes that knowledge integration is a comprehensive process of technology organization and human resource management, in which the initiative and creativity of the integrated entity need to be emphasized. Knowledge integration is an essentially important step in the dynamic process of knowledge innovation.

In the field of computer science and automatic control, knowledge integration research emphasizes on handling organizable and expressible explicit knowledge.

Liu and Ma [14] indicates that, knowledge integration is mainly to identify, process, evaluate and reform new knowledge, to realize interactions between new knowledge and original one, and to provide users with an unified knowledge access interface and intelligent knowledge service by integrating different knowledge structures. Bohlouli et al. [4] investigates a knowledge integration framework based on big data analysis platform, divides knowledge integration processes into acquisition, representation, evaluation, transformation, aggregation and matching of knowledge, which is to provide services for intelligent knowledge retrieval.

In the field of library and information science, relative research is gradually changing from resource integration to resource aggregation. Resource integration refers to combination of all the relative independent resources to a new organic whole, through reorganizing, coordinating, recombining and optimizing the existing status of resource portfolio, which aims to solve the problem of information redundancy, content duplication and inconsistence between primary and secondary documents, while resource aggregation is borrowed from the concept of organic chemistry and refers to fusing knowledge elements to generate new ones by using artificial intelligence technologies, which aims to discover internal semantic associations among resources. Resource aggregation constructs a multidimensional and multi-level resource system with content correlation, and forms a solid knowledge network combining concept themes, subject contents and research objects as a whole [5]. At the conceptual level, KF and resource aggregation have the similar connotations.

Therefore, this paper argues that KF is the advanced stage of knowledge integration. KF applies fusion algorithms and matching rules over the result of knowledge integration to implement deduction, discovery and innovation of knowledge. Furthermore, KF is also difference from knowledge aggregation, in which KF has no need to keep and remain all knowledge concepts, relationships and instances from the original sources, but need to construct the required objects meeting knowledge service demands.

2.3   Fusion of Data, Information and Knowledge

In practice, the term "data", "information" and "knowledge" are not strictly distinguished in statements, and can even be used interchangeably. However, there is a general consensus on distinguishing between the three concepts. A commonly held view, including minor variants is that data is raw numbers and facts without processing, information is processed data, and knowledge is the result of learning and reasoning [1].

The concept of data fusion is mostly in the field of computer science and engineering science. Bleiholder and Naumann [2] indicates that data fusion is the last step in a data integration process, where schemata have been matched and duplicate records have been identified. Data fusion merges duplicate records into a single representation and, at the same time, resolves existing data conflicts. Dong and Gabrilovich [7] also indicates that data fusion aims at resolving conflicts from data and increasing correctness for data integration.

Information fusion is a multidisciplinary research field widely concerned by academic and industrial scientists, and in lots of literature, terms of information/data fusion and information/data integration are used interchangeably. Typically, information fusion refers to the study on efficient methods for automatically or semi-automatically transforming information in time from different sources and different points into a representation that provides effective support for human or automated decision making [3].

Thus, generalized information fusion involves intersections of multiple disciplinary for the processing different information objects. According to application scenarios and processing objects, data/information/knowledge fusions can be regarded as the different levels of abstraction for realizing generalized information fusion. Data fusion is the process of removing noise and redundancy, reducing uncertainty and improving accuracy and reliability of original data at signal and pixel levels. Information fusion is the process of extracting features from multi-source raw data and eliminating contradictions between data contents to improve the consistency and reliability of fused information providing local supports for decision-makers. Data fusion handles raw data on the signal level, and so does information fusion on the feature level. Both of them are belonging to the low-level fusion, while the high-level KF is on the decision level, which involves processes of situation awareness and assessment, influence degree evaluation, fusion optimization, mining implicit information, reasoning and judgment of decision conditions, and so on.

3   Knowledge Representation Based on Ontology

Knowledge representation is the process of symbolizing, formalizing and modeling knowledge, which is the foundation of knowledge organization and the prerequisite for realizing knowledge management. Traditional knowledge representation technologies include state-space, predicate logic, generative rule and frame methods. Along with the discipline crossing and increased complexity of knowledge, methods of neural network, fuzzy set, object-oriented and ontology are developed for knowledge representation. Different knowledge representation methods lead to heterogeneities of knowledge, which is an emerging issue addressed in the research of KF systems.

Although the expressive power and reasoning ability of ontology is less than the traditional formal methods, in order to solve the problem of heterogeneous knowledge, many researches use ontology to represent knowledge and construct knowledge bases [9]. As a structured knowledge representation method, ontology is able to abstractly express a domain as a set of concepts and relationships between the concepts, and unify the domain concepts for sharing the formal specification of the conceptual model, exchanging and reusing knowledge between human and computers. In the Web Ontology Language, *OWL 2*[1], recommended by W3C, the basic modeling elements of ontology are *Classes*, *Properties*, and

---

[1] https://www.w3.org/TR/2012/REC-owl2-primer-20121211/.

Individuals. All entity objects are represented as individuals, while type of entities as classes, and entity relationships as attributes. Attribute can be further refined as sub-attributes, such as object relationships, object features, object value ranges, and so on. Pérez and Benjamins [18] classifies five ontology modeling primitives: *Concepts, Relations, Functions, Axioms* and *Instances.* A concept can be anything including the description of a task, function, action, strategy, reasoning process, etc.; Relations represent a type of interaction between concepts of the domain; Functions are a special case of relations in which the *n-th* element of the relationship is unique for the n-1 preceding elements; Axioms are used to model sentences that are always true; and instances are used to represent elements.

Based on the *OWL 2* definition and Pérez's five modeling primitives, we define a knowledge ontology as the form of five-tuple: $ontology(O) = \langle C, A, R, D, I \rangle$, where C is a set of concepts or classes with hierarchical structure; A is a set of attributes describing features of concepts, and usually defined as attributes of classes; R is a set of relationships, including functions, axioms and other constraints, representing effective associations between concepts, such as father, son and equality relationships, functional relationships and True assertions; D is a set of attribute domains, describing fields or value ranges of attributes; and I is a set of instances, containing entity objects of concept classes.

For example, if $\langle C_H, A_H, R_H, D_H, I_H \rangle$ is defined as an ontology for describing hypertension, set C_H may contain concepts such as $\langle\!\langle HBP \rangle\!\rangle$, $\langle\!\langle Cause \rangle\!\rangle$, $\langle\!\langle Symptom \rangle\!\rangle$, $\langle\!\langle Therapy \rangle\!\rangle$, $\langle\!\langle Patient \rangle\!\rangle$, etc.; set A_H contains attributes of the concepts such as $\langle\!\langle HBP, type \rangle\!\rangle$, $\langle\!\langle HBP, level \rangle\!\rangle$, $\langle\!\langle Cause, humoral \rangle\!\rangle$, $\langle\!\langle Cause, nervous \rangle\!\rangle$, etc.; set R_H indicates relationships between concepts, e.g. $father(\langle\!\langle HBP \rangle\!\rangle, \langle\!\langle PrimaryHBP \rangle\!\rangle)$ means that $\langle\!\langle HBP \rangle\!\rangle$ is the father class of $\langle\!\langle PrimaryHBP \rangle\!\rangle$; and if any, D_H and I_H may contain concept value ranges and its instances.

The five-tuple form reflects the process of hierarchically modeling knowledge from entities to concepts. If only knowledge entities or concepts are separately considered to be merged, the KF process is not comprehensive and completed. In other words, all elements of the knowledge ontology form need to be handled in KF processes, which will be discussed in the next section as KF patterns.

4   Patterns of Knowledge Fusion

So far, there are not many literatures about KF patterns. Xu et al. [24] classifies KF into active and passive types. Qiu and Yu [20] and Zhou et al. [26] discuss several kinds of KF processing algorithms. Smirnov et al. [21] proposes seven context-based KF patterns, i.e. *Simple, Extension, Configured, Instantiated, Flat, Historical* and *Adaptation Fusion*, which are classified upon the problem solved by each KF process for satisfying the requirement of the decision support system.

In this section, we classify KF patterns, from the perspective of knowledge representation, according to the five-tuple ontology form.

Instance Fusion is the process of removing redundancy, deducing noise, correcting error and merging content for entity objects and producing a new set, in which knowledge sources usually have the same modeling structure, or can be converted into the same one. After Instance Fusion, the modeling structure of source knowledge is totally or partly inherited into the fused target in accordance with user definitions and requirements, where the pertinence, consistency and correctness of knowledge entities are improved. There is a substantial overlap between Instance Fusion and traditional information fusion, so that the former can be implemented by using the latter fusing methods as references.

Domain Fusion is the process of applying set operations like UNION, INTERSECT, MINUS and EXCEPT on attribute fields or value ranges of source knowledge entities, resulting in attribute definitions of fused knowledge entities. When Instance Fusion is applied, knowledge sources might be in the same modeling structure but different domains, which is required to redefine the attribute domain of fused knowledge. Domain Fusion remains the modeling structure of source knowledge, but change its attribute fields or value ranges, which is an extension and expansion of Instance Fusion.

Relationship Fusion is the process of merging relationships in source knowledge by removing redundancy and combining structures, as well as applying inductive and deductive reasoning over relationships for inferring and mining a new one. Relationships in knowledge ontology include interactions between concepts, affiliations between concepts and attributes, functions defining particular mappings, and axioms representing true assertions. Relationship Fusion explores and derives new relationships according to original ones in the source, in which modeling structures might be different from either each other, or the fused one where the new knowledge is generated.

Attribute Fusion is the process of comparing, analysing, transforming and merging attributes of knowledge concepts, in terms of classifying, selecting and reorganizing the object features according to users requirements. In the situation of Attribute Fusion, there are usually differences between modeling structures of knowledge sources, especially including complementary, contradiction and homograph differences in attribute definitions. After Attribute Fusion, new attributes appear in the fused knowledge, and new relationships are also required to correspond with them. Thus, Attribute Fusion and Relationship Fusion are two complementary and alternately iterative processes, both are important parts of knowledge discovery and innovation processes.

Concept Fusion is the process of constructing new knowledge concepts, which might bring about new attributes and new relationships as well. Therefore, it is not possible to individually produce Concept Fusion separately from the other KF patterns, which have to be based on Instance Fusion, iteratively and incrementally applying Domain, Relationship and Attribute Fusions to achieve a whole fusion process. Concept Fusion is considered as the high level of the KF hierarchy, where Domain, Relationship and Attribute Fusions are middle levels between the low level Instance Fusion and the high level Concept Fusion. It is difficult to directly apply traditional information fusion methods for Concept

Fusion to generate new knowledge, thus new KF approaches need to be developed, and participations of domain experts are also required for the completion of knowledge innovation.

5   Process Model of Knowledge Fusion

As discussed above, different KF patterns meet different requirements and produce different fusion results. This section proposes two types of process models to analyse the operational mechanism of KF patterns.

5.1   One-Dimension KF Process Model

Relationship, Attribute and Concept Fusions are processes of knowledge innovation, to a certain extent, by changing the original knowledge models and generating a new one; Instance Fusion changes knowledge objects in terms of consistency, correctness, validity and quantities, which is a process of manifesting and discovering knowledge; and Domain Fusion is the transitional phase from knowledge discovery to knowledge innovation, which does not change the original knowledge model but the value range of the concepts.

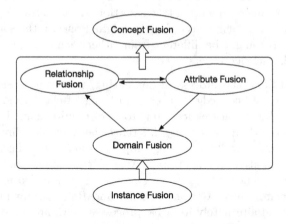

Fig. 1. One-dimension KF process model

Figure 1 gives the one-dimension KF process mode to illustrate relationships among the five KF patterns. The requirement of Domain Fusion is generated on the basis of Instance Fusion. In different knowledge sources, value ranges of concepts might be different from each other, which is required to be adjusted, merged and redefined, i.e. producing Domain Fusion, to meet the demand of Instance Fusion. After changes of concept domains, relationships between the concepts may also need to change so as to affect the inferring results of Relationship Fusion. E.g. the increase or decrease of a concept value ranges is likely to affect the establishment of equal relationships between the concepts. At the same time, Relationship Fusion and Attribute Fusion are also two interactive and complementary processes. The production of new attributes might lead to the generation of new relationships, and vice versa.

Therefore, the three KF patterns, i.e. Domain Fusion, Relationship Fusion and Attribute Fusion, are performing in a way of loop iterations. In order to eventually achieve Concept Fusion, each iteration makes a further step in the progress of generating new knowledge. Thus, KF processes could not be completed only by a single fusion pattern, nor by a stepwise linear procedure. All fusion patterns need to be comprehensively considered, and KF is realized in a way of loop iteration, incremental progression and spiral development.

5.2 Two-Dimension KF Process Model

As mentioned above, KF generates new knowledge and produces knowledge innovation, while the aim of knowledge innovation is to provide better knowledge service. Nonaka et al. [17] summarizes knowledge innovation processes into four stages: *Socialization, Externalization, Combination* and *Internalization,* as known as the SECI model, describing transformations between tacit and explicit knowledge. Socialization is the process of converting new tacit knowledge through shared experiences; Externalization is the process of articulating tacit knowledge into explicit knowledge; Combination is the process of converting explicit knowledge into more complex and systematic sets; Internalization is the process of embodying explicit knowledge into tacit knowledge.

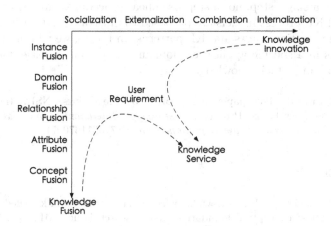

Fig. 2. Two-dimension KF process model

In the SECI model, knowledge is created through a spiral by applying the four processes in a way of circular loop rather than a stepwise linear procedure, which is similar to the implementation of KF patterns. Although it is not able to directly map the KF patterns with the SECI stages, the common characteristic makes it possible to organically combine the two processes accordingly, as shown in Fig. 2, in order to achieve the accurate, personalized and effective knowledge service in accordance with the user requirement. In particular, during the stages of Socialization and Externalization, methods for fusing instances and domain can be used to discover tacit knowledge objects, and methods for fusing relationships and attributes can be used to articulate it into an explicit one, while during the stages of Combination and Internalization, the fusion patterns are naturally involved since they are both supposed to handle explicit knowledge.

The two-dimensional KF process model shows relationships between the innovation stages and the fusion patterns and indicates that, although KF patterns proposed in this paper are based on the ontology representation of explicit knowledge, it have the potential to expand to tacit KF, which is one of the research issues in our future work.

6   Conclusion and Future Work

The big data era brings distributed, heterogeneous and autonomous knowledge, from which KF integrates, discovers and exploits valuable knowledge for achieving a high quality service. This paper discuss the KF connotation in terms of giving the definition of KF and analysing the relation and difference between KF and various notions, such as knowledge integration, information fusion and data fusion. Then, we introduce five KF patterns, i.e. *Instance*, *Domain*, *Relationship*, *Attribute* and *Concept Fusion*, and indicate that the KF process is implemented in a way of loop iteration, incremental progression and spiral development, rather than only by a single step, nor a stepwise linear procedure. Finally, two types of dimensional KF process models are proposed to illustrate relationships between knowledge innovation stages and KF patterns. In future, we will implement the KF patterns in a specific application domain, e.g. chronic disease domain, and extend it to handle tacit knowledge.

Acknowledgement. This paper is supported by the Chinese NSFC International Cooperation and Exchange Program, *Research on Intelligent Home Care Platform based on Chronic Diseases Knowledge Management* (71661167007).

References

1. Alavi, M., Leidner, D.E.: Review: knowledge management and knowledge management systems: conceptual foundations and research issues. MIS Q. **25**, 107–136 (2001)
2. Bleiholder, J., Naumann, F.: Data fusion. ACM Comput. Surv. **41**(1), 1–41 (2008)
3. Balazs, J.A., Velasquez, J.D.: Opinion mining and information fusion: a survey. Inf. Fusion **27**, 95–110 (2016)
4. Bohlouli, M., Merges, F., Fathi, M.: Knowledge integration of distributed enterprises using cloud based big data analytics. In: Proceedings of IEEE International Conference on Electro/Information Technology, pp. 612–617, 5–7 June 2014

5. Bi, Q.: Digital resources: from integration to aggregation. Digit. Libr. Forum **6**, 1 (2014)
6. Cai, Q.H., Chen, G.H.: A review of knowledge integration research. J. Res. Dev. Manag. **22**(6), 15–22 (2010)
7. Dong, X.L., Gabrilovich, E.: From data fusion to knowledge fusion. In: Proceedings of VLDB 2014 (2014)
8. Dong, X.L., Srivastava, D.: Knowledge curation and knowledge fusion. In: Proceedings of VLDB, pp. 2063–2066 (2015)
9. Guo, Q., Guan, X., Cao, X.Y., et al.: Research progress and trends of knowledge fusion. J. China Acad. Electron. Inf. Technol. **7**(3), 252–257 (2012)
10. Hu, S.K., Cao, Y.D.: Knowledge fusion framework based on web page texts. Front. Comput. Sci. China **3**(4), 457–464 (2009)
11. Hou, J., Yang, J.G., Jiang, Y.L.: Knowledge fusion algorithm based on metadata and ontology. J. Comput.-Aided Des. Comput. Graph. **18**(6), 819–823 (2006)
12. Kampis, G., Lukowicz, P.: Collaborative knowledge fusion by ad-hoc information distribution in crowds. Proc. Comput. Sci. **51**, 542–551 (2015)
13. Liu, X.C., An, X.M.: Knowledge integration research status analysis. Inf. Doc. Serv. **1**, 9–12 (2006)
14. Liu, X.L., Ma, J.: Research progress of knowledge integration based on Ontology in semantic web environment. J. Modern Intell. **01**, 159–163+169 (2015)
15. Liu, J., Xu, W., Jiang, H.: Research on dynamic ontology construction method for knowledge fusion in group corporation. In: Wen, Z., Li, T. (eds.) ISKE 2013. AISC, vol. 278, pp. 289–298. Springer, Heidelberg (2014). doi:10.1007/978-3-642-54930-4_29
16. Meng, X.F., Chi, X.: Big data management: concepts, technologies and challenges. Comput. Res. Dev. **50**(1), 146–169 (2013)
17. Nonaka, I., Umemoto, K., Senoo, D.: From information processing to knowledge creation: a paradigm shift in business management. Technol. Soc. **18**(2), 203–218 (1996)
18. Pérez, A.G., Benjamins, V.R.: Overview of knowledge sharing and reuse components: ontologies and problem-solving methods. In: Proceedings of the IJCAI-1999 Workshop on Ontologies and Problem-Solving Methods (KRR5) (1999)
19. Preece, K., Hui, A.G., et al.: Kraft: An agent architecture for knowledge fusion. Int. J. Coop. Inf. Syst. **10**(1–2), 171–195 (2001)
20. Qiu, J.P., Yu, H.Q.: Research progress and trends of knowledge fusion in perspectives of knowledge science. Libr. Inf. Serv. **59**(08), 126–132+148 (2015)
21. Smirnov, A., Levashova, T., Shilov, N.: Patterns for context-based knowledge fusion in decision support systems. Inf. Fusion **21**, 114–129 (2015)
22. Suchanek, F.M., Weikum, G.: Knowledge bases in the age of big data analytics. In: Proceedings of VLDB Endowment, vol. 7, pp. 1713–1714 (2014)
23. Tang, X.B., Wei, W.: The growth points of knowledge service in big data age. Res. Libr. Sci. **05**, 9–14 (2015)
24. Xu, C.J., Li, A.P., Liu, X.M.: Knowledge fusion architecture. J. Comput.-Aided Des. Comput. Graph. **22**(7), 1230–1236 (2010)
25. Ye, Y., Ma, F.C.: The rise of data science and its relation with information science. J. Inf. Sci. **34**(6), 575–580 (2015)
26. Zhou, F., Wang, P.B., Han, L.Y.: Multi source knowledge fusion processing algorithm. J. Beijing Univ. Aeronaut. Astronaut. **39**(1), 109–114 (2013)

E-SKB: A Semantic Knowledge Base for Emergency

Chang Wen, Yu Liu[✉], Jinguang Gu, Jing Chen, and Yingping Zhang

College of Computer Science and Technology,
Wuhan University of Science and Technology, Wuhan, China
liuyu@wust.edu.cn

Abstract. Although the number of knowledge bases in Linked Open Data has grown explosively, there are few knowledge bases about emergency, an important issue in the area of social management. In this paper, we introduce a semantic knowledge base of emergency, extracted from an authoritative website. According to the characteristics of the website, a framework is suggested to convert web into RDF. In order to help researchers acquire more knowledge, we follow the publishing rules of Linked Open Data—not only using URIs to label the objects in the semantic knowledge base, but also providing links to DBpedia. Finally, we employ Sesame to store and publish the semantic knowledge base, and develop a query interface to retrieve the knowledge base with SPARQL.

Keywords: Emergency · Linked open data · Semantic knowledge base · SPARQL

1 Introduction

Emergency, an unexpected event, may cause serious social harm and bring a great loss to human life [1]. It can be divided into four categories, natural disasters, accidents disasters, public health and social security [2]. Due to the uncertainty and paroxysm of emergency, it is necessary for us to integrate the scattered data into a knowledge base, which contributes to collect information efficiently and conveniently.

As one of the widely used technologies, the semantic knowledge base is suitable for developing an application to deal with emergencies. Some popular knowledge bases were constructed successfully such as GeoNames [3], DBpedia [4], FOAF [5], etc. If the knowledge base about emergency has collected a great amount of events in detail, it would be easy to work out potential results and feasible solutions by searching the knowledge base when an emergency happens. The E-SKB can be adopted to construct an expert system to handle emergencies, improve the query accuracy and realize the linguistic diversity through linking with DBpedia.

Since the most of knowledge bases in Linked Open Data (LOD) [6] do not cover the specific knowledge about emergency, we extract the web information

© Springer Nature Singapore Pte Ltd. 2016
H. Chen et al. (Eds.): CCKS 2016, CCIS 650, pp. 196–202, 2016.
DOI: 10.1007/978-981-10-3168-7_19

from a case database of emergency management, maintained by Jinan University, which has collected 574 cases and 2275 resources of emergency (http://decm. jnu.edu.cn/) and can be accessed by the browser. Then the web information is converted into RDF triples by following the principles of LOD, so that researchers can retrieve the knowledge with semantic web technologies, such as SPARQL.

2   Construction Process of E-SKB

In order to construct the E-SKB, the main processing procedures can be divided into three parts. First of all, we introduce a crawler that is applied to collect data from the web. Secondly, we extract the concepts according to the classification tree, then define the properties by the labels on the news pages, and link the data set to DBpedia following the LOD rules. Finally, we employ the Sesame to store the data and develop a query interface to acquire E-SKB by SPARQL.

2.1   Extracting Data from Web

Given there are a wide variety of methods to develop a crawler, we just give a brief description of the crawler that is used by JSOUP and HttpClient.

Due to a large quantity of URLs need to be solved, the technologies of queue and multithreading are applied to the crawler. As a result, the crawling process can be functioned more efficiently. While an URL is added into the queue, we collect the news information with HTML filter and convert it to a JSON string. Then the first URL will be removed and we repeat the previous step until the queue is empty. Finally, the emergency information is presented in the JSON format.

2.2   Processing Data with Certain Rules

The Linked Data is a group of best practices for publishing and interlinking structured data on the web. It was introduced by Tim Berners-Lee in his website [7] and has become known as the Linked Data principles, which can be concluded as follows:

- Using URIs to present things.
- Using HTTP URIs, so that people have access to resources.
- When someone looks up the URI, information is found by SPARQL query language.
- URIs are linked with each other, helping users discover more resources.

In the process of dealing with the data set, the principles mentioned above should be obeyed in order that we can keep the data normatively.

Concepts Extraction. Concepts are utilized to describe a set of entities, which possess the same types and can be linked to the existed ones on the Internet. The emergencies are classified into several kinds based on the classified layer tree on the web. Figure 1 shows some event classes and the relations between them, the "Accident Disaster" can be regarded as a concept that has nine sub classes, each one of them is a unique concept as well. All the entities belong to the nine sub classes are parts of the "Accident Disaster". The main relation of these concepts is presented by the property "subClassOf" in RDFS, which means one concept is a subset of another.

According to the Linked Data principles, each concept has a unique URI so that there is no confliction in defining the emergencies. We construct the concept's URI by adding the class name behind the namespace. The class names are extracted from the classified layer tree in the home page, and the namespace is defined as "http://decm.jnu.edu.cn/class#". Finally, we can build a concept model to show the taxonomic hierarchies between different emergencies.

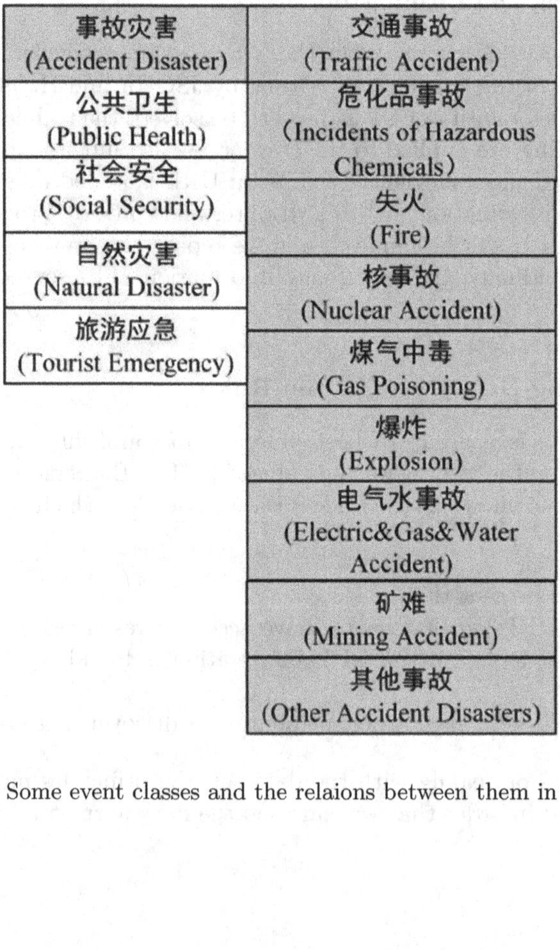

事故灾害 (Accident Disaster)	交通事故 (Traffic Accident)
公共卫生 (Public Health)	危化品事故 (Incidents of Hazardous Chemicals)
社会安全 (Social Security)	失火 (Fire)
自然灾害 (Natural Disaster)	核事故 (Nuclear Accident)
旅游应急 (Tourist Emergency)	煤气中毒 (Gas Poisoning)
	爆炸 (Explosion)
	电气水事故 (Electric&Gas&Water Accident)
	矿难 (Mining Accident)
	其他事故 (Other Accident Disasters)

Fig. 1. Some event classes and the relaions between them in E-SKB

Properties Extraction. The properties are relations between the subject resources and object resources, which can be deemed to the predicates in the sentences. The properties are divided into two groups, the system properties and user defined properties. System properties are the internal properties of the RDF and RDFs, which have XML Schema data type vlaues. User defined properties are the attributes defined to present the specific relations. For each defined properties, we need to assign its domain and range to indicate the subject and object.

The definition of property is the same as the concept, which includes namespace and property names. The namespace is defined as "http://decm.jnu.edu.cn/property#". Since most of the pages are constituted in a uniform way, all of them include the same labels to present the emergency contents. Figure 2 shows an emergency case entitled "Spraying pesticide poisoned 9 people". The seven labels are nation, area, location, start time, end time, loss and relevant resources. We regard these labels as the property names and define a "content" to present the description. So the "content" is expressed as "http://decm.jnu.edu.cn/property#content", which is abbreviated to "depr:content".

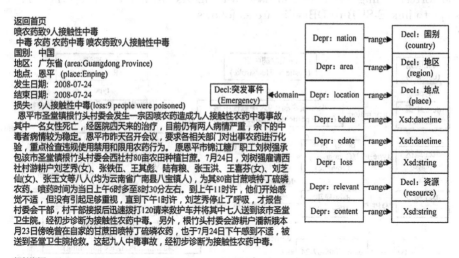

Fig. 2. The properties of an event instance

Instances Extraction. According to the definitions of concepts and properties, we extract the instances from the web. They can be divided into two types, emergency news and related resources. The former is the news that is described in the page, and the latter is related news about the topic. The relation between two instances is represented as the property "depr:relevant" as we have mentioned above.

Take the news of "Spraying pesticide poisoned 9 people" for example, the knowledge graph of the instance is shown in Fig. 3. Resources are connected with the instance by properties: therefore they constitute triples that can be formatted into RDF.

2.3 Linking E-SKB to DBpedia

Linked Data is the core technology in exposing, sharing and connecting web information, which uses RDF and URI to present things and the relations between resources. The characteristics of LOD include simple structures, standardized information and low-cost interaction between the mankind and the machine. In this paper, the geographical concepts can be associated with the resources in the DBpedia for the sake of data sharing.

As mentioned before, the instances have the properties of nation, area and location, and the property values are the resources of the specific information. We can connect these geographical values with the resources that have the same meanings in DBpedia. Since the values in E-SKB are Chinese, we need to find the corresponding resources and use "owl:sameAs" to link them together. The steps to link E-SKB to DBpedia are as follows:

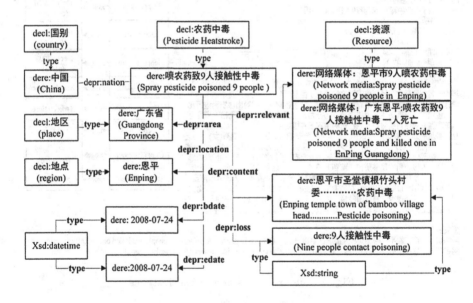

Fig. 3. The knowledge graph of an event instance

1. Constructing the geographical resources according to the format in DBpedia. The prefix of resources is defined as "http://dbpedia.org/resource/", so we can add the Chinese geographical names after the prefix to construct the resources.

2. Querying the related resources by using the SPARQL endpoint in DBpedia. We use the resources built in step one as the objects and the "http://dbpedia.org/ontology/wikiPageRedirects" as the predicate to construct the query statements. The related resources in DBpedia such as "http://dbpedia.org/resource/Beijing" will be returned.
3. Linking the geographical resources to DBpedia. We use "http://www.w3.org/2002/07/owl#sameAs" as the property to link E-SKB to the resources returned in step two.

According to the steps discussed above, we can get 891 linking results of the geographical resources in E-SKB. In the future work, we will expand the E-SKB by extracting other news websites and linking more resources to DBpedia.

3 Publishing E-SKB into Sesame

We store the data as RDF triples and publish it into the Sesame server. The Sesame files are downloaded and deployed to the tomcat server. Finally, the RDF file is uploaded to the Sesame server, which can be accessed by the query interface.

4 Web-Based Query System

In the web-based query system, we can get the detailed information of E-SKB by SPARQL. Figure 4 shows the result of querying instance "Spraying pesticide poisoned 9 people", the properties and objects are returned.

http://www.w3.org/1999/02/22-rdf-syntax-ns#type	http://decm.jnu.edu.cn/class#农药中毒
http://decm.jnu.edu.cn/property#Relations	http://decm.jnu.edu.cn/class#网络媒体: 广东恩平:喷农药致9人接触性中毒 一人死亡
http://decm.jnu.edu.cn/property#Areas	http://decm.jnu.edu.cn/class#广东省
http://decm.jnu.edu.cn/property#StartTime	http://decm.jnu.edu.cn/propery#2008-07-24
http://decm.jnu.edu.cn/property#Relations	http://decm.jnu.edu.cn/class#网络媒体: 恩平市9人喷农药中毒
http://decm.jnu.edu.cn/property#Damages	http://decm.jnu.edu.cn/propery#9人接触性中毒
http://decm.jnu.edu.cn/property#Countrys	http://decm.jnu.edu.cn/class#中国
http://decm.jnu.edu.cn/property#EndTime	http://decm.jnu.edu.cn/propery#2008-07-24

Fig. 4. Query results of an event instance

Acknowledgments. This work was partly supported by the National Science Foundation of China (No. 61502359), the National Students' Innovative Entrepreneurship Training Program under Grant (No. 201510488016).

References

1. Lan, X., Kaibin, Z.: The category, classification, and periodization of emergency events: the based management system of emergency. Administrative Management of China, 102–107 (2005)
2. An, Y.: The theoretical framework of emergency response law. Law Sci. Mag. **27**(4), 28–31 (2006)
3. Yoshioka, M., Kando, N.: Issues for linking geographical open data of geonames and wikipedia. In: Takeda, H., Qu, Y., Mizoguchi, R., Kitamura, Y. (eds.) JIST 2012. LNCS, vol. 7774, pp. 375–381. Springer, Heidelberg (2013). doi:10.1007/978-3-642-37996-3_32
4. Auer, S., Bizer, C., Kobilarov, G., et al. DBpedia: A Nucleus for a Web of Open Data (2010)
5. Dan, B.L.: FOAF vocabulary specification 0.9. Comput. Sci. Commun. Dictionary **23**(3), 165 (2007)
6. Heath, T., Bizer, C.: Data: evolving the web into a global data space. Molecular Ecol. **22**(3), 670–684 (2011)
7. Tim Berners-Lee. Linked Open Data Design Issues. http://www.w3.org/DesignIssues/LinkedData.html (2006)

CCKS 2016 Shared Tasks

ICRC-DSEDL: A Film Named Entity Discovery and Linking System Based on Knowledge Bases

YaHui Zhao, Haodi Li, Qingcai Chen, Jianglu Hu, Guangpeng Zhang,
Dong Huang, and Buzhou Tang[✉]

Harbin Institute of Technology, 518000 Shenzhen, China
yahuizhao.hitsz@gmail.com, haodili.hit@gmail.com,
qingcai.chen@gmail.com, hujianglu.hit@gmail.com,
donghuang2010@gmail.com, tangbuzhou@gmail.com,
zhangguangpeng_hit@163.com

Abstract. Named entity discovery and linking are hot topics in text mining, which is very important for text understanding as named entities that usually presented in various formats and some of them are ambiguous. To accelerate the development of related technology, the China Conference on Knowledge Graph and Semantic Computing (CCKS) in 2016 launches a competition, which includes a task on film named entity discovery and linking (i.e., task 1). We participate this competition and develop a system for task 1 of the CCKS competition. The system consists of two individual parts for named entity discovery (NED) and entity linking (EL) respectively. The first part is a hybrid subsystem based on conditional random field (CRF) and structural support vector machine (SSVM) with rich features, and the second part is a ranking subsystem where not only the given knowledge base but also open knowledge bases are used for candidate generation and SVM^{rank} is used for candidate ranking. On the official test dataset of Task1 of CCKS 2016 competition, our system achieves an F1-score of 77.83% on NED, an accuracy of 86.53% on EL and an overall F1-score of 67.35%.

Keywords: Entity recognition · Entity linking · CRF · Learning to rank

1 Introduction

Named entity discovery (or recognition) (NED) and linking (EL) are two fundamental tasks of text mining, which play an important role in many applications such as question answering, information retrieval, etc. NED as a traditional problem of natural language processing has been comprehensively studied for a long time. However, most studies focus on NED in the newswire domain, NED in a new specific domain is still a challenge. EL is one of key steps to understand text when given one or more knowledge bases (KBs). It is also a significant research topic of knowledge graph, which has attracted considerable attention in recent years. In the past several years, a small number of methods have been proposed for EL, however, it is not easy to evaluate which one is better than another one as there are few publicly available corpora.

© Springer Nature Singapore Pte Ltd. 2016
H. Chen et al. (Eds.): CCKS 2016, CCIS 650, pp. 205–213, 2016.
DOI: 10.1007/978-981-10-3168-7_20

To accelerate the development of related research, the China Conference on Knowledge Graph and Semantic Computing (CCKS) in 2016 organizes a competition, which includes a task on film NED and NEL (i.e., task 1), which is domain-specific. The goal of NED is to find out film-related entities, such as film names, directors and actors, from public film reviews, and NEL is to link the extracted entities to a given KB.

We participate this competition and developed a pipeline system for task 1 of this competition. In our system, we treat NED as a sequence labeling problem and EL as a ranking problem. Conditional random field (CRF) and structural support vector machine (SSVM) are employed for NED, and SVM^{rank} for EL. We developed our system on the training set of the competition, and evaluation on an independent test set shows that the F1-score of our system on NED is 77.83%, accuracy on EL is 86.53% on EL and overall F1-score is 67.35%, which is competitive other participating systems.

2 Related Work

From the beginning of the last century, Message Understanding Conference (MUC), Automatic Content Extraction (ACE), Multilingual Entity Task (MET) and some other conferences have held ongoing. The research of Information Extraction has been developed and popularized.

Information understanding conference has an important role in promoting the research of information processing, named entity discovery or recognition as a task to be researched, can be traced back to the 7th IEEE artificial intelligence application conference in 1991. In this conference, Ran published a paper about "extract and recognize company name" [4]. Ran introduced a system that can recognize and extract company name, it based on rule and heuristic algorithm. In 1996, named entity recognition as a subtask of information extraction formally introduced to MUC-6. On MUC- 6, named entity was defined included: Person, Location, Organization, Date, Time, Percentage, and Monetary Value. On MUV-7 [5], information extraction was defined included three tasks as follows: Template Element(TE), Template Relation (TR), Scenario Template(ST). Unfortunately, few studies have been carried out on the film-related entity discovery in Chinese.

The input of Entity Linking (EL) is a mention in a text [6], and the task of EL is to make the mention link to a certain entity in the specified knowledge base. Generally, The task of EL include two main steps [7], candidate entity generation and candidate entity ranking. Candidate entity generation aims to find all possible entities of entity mention in knowledge base. Then extracting features from these entities and ranking, retrieve the optimal results by (re-)ranking methods at last. In recent years, there has been an increasing amount of literature on EL. Name dictionary based techniques and search engines based methods are most common used in candidate entity generation while series methods such as learning to rank, graph based, vector space model, etc. are popular used in candidate entity ranking. Although many research has been carried out on the general domain such as Wikipedia, few studies focused on the special domain. Moreover, to the best of our knowledge, previous studies of entity linking have not dealt with film-related entities in Chinese.

3 Methods

The Task1 of CCKS 2016 consists of two subtasks: NED and EL. According to the amount of entities existed in the assigned KG, it is prone to linking non-entity mention to KG in error when using joint learning method. Therefore, our approach of this competition separate process the two tasks with two independency module as a pipeline system but not used a jointly methods.

In the entity discovery task, we first extracted the features that contain the matching of entity knowledge base, pinyin of characters, part of speech of words, sentence information extract by deep learning method and so on. Additional, we use structured SVMs [9] as an auxiliary methods predict an independence result, and then merge them by simple combination.

In the entity linking task, we use a hierarchical filter so called *sieve-based filter* to achieve candidate entities of each mention, and then we retrieve the most relevant entity by ranking those candidates used learning to rank method. Its overall process as shown in Fig. 1 below:

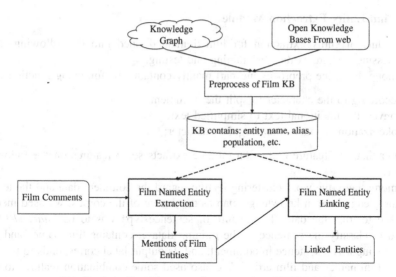

Fig. 1. Overall architecture of our system: ICRC- DSEDL

First, we analyse and rebuild the film-related knowledge base by extracting the key information for the tasks from the assigned KG and the open knowledge base from web.

We were obtaining the knowledge information of film related entity names, alias, and popularity and so on. Based on the knowledge information, we extracted the mentions of film artists and film names through entity extraction module. After that, the film-related entity linking module links the extracted mentions to the assigned KG, the results of entity discovery and linking obtained.

3.1   Rebuild Knowledge Base

Firstly, this module transforms the assigned KB to structured data, extracting the film name and its related information such as director name, actor name. Secondly constructing interface according to the film name to return the information about the film. Besides, we also rebuild the knowledge base by adding two parts of information of entities:

(1) Entity popularity. Through the open application program interface of Douban website getting popularity information, including the number of film works, the number of comments on film works, the number of collection.
(2) Entity alias. From the open knowledge base source such as Baidu, Douban and Wikipedia, we constructing the film work alias dictionary (such as the name of a foreign language, the translation name of Hong Kong and Taiwan, Abbreviation, etc.), and the film artist alias dictionary (such as foreign name, Chinese translation name, common nickname, etc.).

3.2   Film Entity Extraction Module

The module of entity extraction for film domain divided into the following steps: preprocessing, feature extraction, training and testing.

Among them, the preprocessing part mainly contains the following functions:

(1) According to the character to split the document;
(2) Convert the traditional text to simplified text;
(3) Tokenization, standard the irregular character;

After that, the feature extraction module extracts some features on the following Table 1.

Among them, the word clustering feature regard the comment data and the text of film knowledge base in knowledge map as the source of the corpus, we used method was used to train the data. The feature of sentence type was to use *stacked-LSTM* method to classify the sentence in the training dataset contains film names and film artists. Its input is a sentence in comment, and its output label corresponds to whether existing film names and film artists. We also used some combination features to help abundance our features. The combination features we used contain "Character + Word boundary", "Character + Part of speech", "Part of speech + Word boundary" and "Character + Sentence type". We used BIOES tagging rules in the sequence labeling system.

3.3   Film Entity Linking Module

The module of entity linking divided into two steps: (1) candidate entity generation; (2) candidate entity ranking.

Table 1. Feature list of film entity recognition module

Feature name	Feature type	Feature description
Character	Text features	A single character, and N-grams (N = 1,2,3,4)
Pinyin		Pinyin corresponding to each character
Word boundary		The boundary of the word where the character is located
Part of speech		The part of speech of the word where the character is located
Word clustering		Clustering features of word vectors obtained by word embedding method
The source of sentence		Depending on the source of the sentence, used to distinguish between different types of comments
Sentence type		Using RNN to learn and determine whether the sentence may describe the film artist or film work
Film artist	Knowledge features	Whether to match with the film artist dictionary in the knowledge base
Film work		Whether to match with the film work dictionary in the knowledge base
Related film artist		According to the title of the comment, obtaining a list of the related film artist, and determining whether matching the list
Related film work		According to the title of the comment, obtaining a list of the related film work, and determining whether matching the list
Alias		Whether matching the alias dictionary of knowledge base
Popularity		The popularity of the film artist and the film name

The generation of candidate entity use *sieved* strategy rather than one-off, we extract candidates gradually follow the level as following, if the candidate set in an upper level is not empty, stop the extraction.

(1) Exact matching, for entries in the knowledge base that can exact match with mention, put these entries into candidate set.

(2) Partial matching, for the entries in the knowledge base that can partial match with mention, put these entries into candidate set.

(3) All-word-in matching, if the name of entries contains all the words of mention, put these entries into candidate set.

(4) Edit distance matching, first set up a threshold if the length of mention is shorter than four, otherwise is two. Then calculate the edit distance between entity mention and entry in knowledge base. Put entries that edit distance smaller than threshold.

(5) Pinyin edit distance matching, set up a threshold if the length of mention is shorter than four, otherwise is two. Then calculate the pinyin edit distance between entity mention and entry in knowledge base. Put entries which pinyin edit distance smaller than threshold.

After the candidate entity generation, we need to rank these candidate entities. Our system uses supervised ranking methods, the features we used as follows.

The first feature we used is name string comparison. For the entity in candidate set, calculate the edit distances between full name, alias, pinyin of full name, pinyin of alias in knowledge bases with extracted mentions, and used the smallest one as the edit distance feature. The second feature is entity popularity. We can get ratings counts of film entity and fans counts of celebrity entity in Douban. Ratings counts show how many people rat this film, fans counts show how many people are fans of this celebrity. For a film entity, the formula of entity popularity as follows:

$$P_m(e) = \frac{Ratings(e)}{\sum_{k=1}^{n} Ratings(e_k)}$$

Where e is a candidate entity, $Ratings(e)$ is the ratings count of e, n is the length of candidate entity. For a celebrity entity, the formula of entity popularity as follows:

$$P_h(e) = \frac{Fans(e)}{\sum_{k=1}^{n} Fans(e_k)}$$

Where e is a candidate entity, $Fans(e)$ is the fans counts of e, n is the length of candidate entity.

Similarity based on keywords is another feature, for each film entity, we can get the keywords information from Douban, for example, the key word of film *Jaws* is "thriller, US, disaster, classic, terror, 1975, plot, science fiction". We use the brief introduction of celebrity as their key word. For an entity e, the formula of keyword similarity as follows:

$$Sim(e, K) = \frac{counts(e)}{length(k)}$$

Where e is an entity, K is a keyword, $counts(e)$ is the number of keyword occurrences in a review, $length(K)$ is the number of keywords.

Last, we use related feature that we get a list of all actors and directors of the film about each review. If the candidate entity in the list, set related feature 1, otherwise set 0.

4 Experimental Result

Keg-Movie-Ontology (KMO) is a completely structured bilingual film KG constructed by department of computer knowledge engineering lab of Tsinghua University, including 23 concepts, 91 attributes, more than 700,000 entities and more than 10 million triple. This task used a subset of KMO, which only include entities of Douban as the assigned KG.

KMO includes the following documents: (1) the artist entity contains artist name and personal information (2) the film entity contains film name, actor, director, producer, etc. (3) the concept and hyponymy (4) the actor information in filmography.

We resolve the knowledge mapping to linked data format by data extraction from KMO. We extracted film name, director name, artist name, construct API of director and artist information.

We treat NED as a sequence labeling problem, and the sequence annotation problem is a typical problem in Natural Language Processing domain. CRFs is one of the most common model used to solve sequence tagging problem, therefore, we use CRFs [8] as our main method to recognize entity. Additional, we use structured SVMs [9] as an auxiliary methods predict an independence result, and then merge them by simple combination.

Based on the training set and test set published by organizer, we have carried out an experimental evaluation of the entity recognition system. We use the training data to train the model, and use the test data to test the performance of the model. Entity recognition uses Precision, Recall and F1-Measure as evaluation indicators. The results of the experiments shown in Table 2:

Table 2. Entity recognition performance of ICRC-DSEDL system

Method	Features	Precision (%)	Recall (%)	F1 (%)
CRFs	N-Grams	79.67	38.74	52.13
	N-Grams + character text features	77.86	59.81	67.65
	N-Grams + character text features + sentence source	80.49	58.66	67.86
	N-Grams + character text features + sentence source + RNN	80.33	58.85	67.93
	N-Grams + character text features + sentence source + RNN + knowledge features	84.37	69.92	76.47
CRFs + Structural SVM	N-Grams + character text features + sentence source + RNN + knowledge features	82.10	73.98	**77.83**

As the Table 2 shows, the character text features are more significant for NED tasks, because the word boundaries and part of speech after the word segmentation play a great role. The knowledge base also has a great influence on the entity discovery module, it increase 8.54 percent point on the foundation of text features. Besides, multi-model combination can improve the performance, because each model just identifies the limited entities.

The entity linking regarded as a ranking problem in our system. First, find the candidate entity in the give knowledge base, then rank them, and finally obtain the top ranked as the linking entity. Ranking SVM employed to rank the candidate in the EL module. We use edit distance, pinyin edit distance, popularity and related features to decide the rank result of candidate entity.

In the part of EL, we use accuracy as evaluation. The results of the experiments are shown in Table 3, the mentions need linked were extracted by our NED module and there are five different type of film comments, we demonstrate the accuracy (%) by adding features in each type. The edit distance, popular and keyword sim in Table 3 are indicating features of pinyin edit distance, entity popularity and similarity based on keywords, respectively.

Table 3. Entity linking performance of ICRC-DSEDL system

Dataset	Edit distance	Edit distance + popular	Edit distance + popular + keyword sim
Group topic	84.16	89.15	85.71
Review	79.58	78.85	88.67
Review comment	72.64	84.91	82.58
Short comment	72.46	80.84	78.10
Synthetic review	81.25	87.50	92.00
Total	79.41	82.22	**86.53**

The edit distance can partly provide distinction between candidate entities, however, because of the ambiguity of entity, we need more information to distinguish different candidate. Popularity is a very useful feature, and it shows popularity of entities with same name. In addition, keyword similarity can bring context information to contribute candidate entity ranking.

At last, we evaluate both NED and EL, on the level of end to end, to make a comprehensive evaluation of the entire system. The evaluation results shown in Table 4, each.

Table 4. An end to end comprehensive evaluation of ICRC-DSEDL system

Dataset	NED (%)			EL (%)	Overall (%)		
	P	R	F1	Accuracy	P	R	F1
Group topic	84.29	73.48	78.52	85.71	72.24	62.98	67.29
Review	81.22	70.35	75.39	88.67	72.02	62.38	66.85
Review comment	76.40	85.00	80.47	82.57	63.08	70.18	66.44
Short comment	84.23	85.21	84.72	78.09	65.78	66.54	66.16
Synthetic review	86.32	82.79	84.52	92.00	79.41	76.17	77.76
Total	82.10	73.98	**77.83**	**86.53**	71.04	64.01	**67.35**

5   Conclusion

This paper introduces build a film-related named entity discovery and linking pipeline system. It has two modules and includes the three-part works: rebuild knowledge base, NED module and EL module. First, we work on a series of data extraction and analysis, and then transform the KG to structured related data. After that, we look through open knowledge bases from web resource to enrich the knowledge information of our knowledge base. Then, we use CRFs as our main method to recognize entity, and structured SVMs as auxiliary for the entity discovery subtask. Finally, we use a sieve-based method to get candidate entities for extracted mentions and obtain the anchor entity by ranking SVM method.

Acknowledgments. This paper is supported in part by grants: National 863 Program of China (2015AA015405), NSFCs (National Natural Science Foundation of China) (61402128, 61473101, 61173075 and 61272383) and Strategic Emerging Industry Development Special Funds of Shenzhen (JCYJ20140508161040764, JCYJ20140417172417105 and JCYJ20140627163809422)

References

1. Santos, C.N., Milidiú, R.L.: Named entity recognition. Entropy Guid. Transform. Learn. Algorithms Appl. 51–58 (2012)
2. Nanyun, P., Dredze, M.: Named entity recognition for chinese social media with jointly trained embeddings. In: Proceedings of EMNLP (2015)
3. Li, H.: Learning to rank for information retrieval and natural language processing. Synth. Lect. Hum. Lang. Technol. **7**(3), 1–121 (2014)
4. Grishman, R., Sundheim, B.: Message understanding conference-6: a brief history. In: COLING, vol. 96, pp. 466–471 (1996)
5. Chinchor, N., Marsh, E.: Muc-7 information extraction task definition. In: Proceeding of the Seventh Message Understanding Conference (MUC-7), Appendices, pp. 359–367 (1998)
6. Shen, W., Wang, J., Han, J.: Entity linking with a knowledge base: issues, techniques, and solutions. IEEE Trans. Knowl. Data Eng. **27**(2), 443–460 (2015)
7. Yuan, J., Yang, Y., Jia, Z., Yin, H., Huang, J., Zhu, J.: Entity recognition and linking in Chinese search queries. In: Li, J., Ji, H., Zhao, D., Feng, Y. (eds.) NLPCC 2015. LNCS (LNAI), vol. 9362, pp. 507–519. Springer, Heidelberg (2015). doi:10.1007/978-3-319-25207-0_47
8. Lafferty, J., McCallum, A., Pereira, F.: Conditional random fields: probabilistic models for segmenting and labeling sequence data. Dep. Pap. CIS, June 2001
9. Altun, Y., Tsochantaridis, I., Hofmann, T.: Hidden Markov support vector machines. In: ICML 2003, vol. 3, pp. 3–10 (2003)

Domain-Specific Entity Discovery and Linking Task

Tao Yang, Feng Zhang$^{(\boxtimes)}$, Xiao Li, Qianghuai Jia, and Ce Wang

Tencent Inc., Beijing, China
{rigorosyang, jayzhang, chinali,
jasonqhjia, fordwang}@tencent.com

Abstract. This paper describes the TEDL system for the entity discovery and linking, which compete the CCKS2016 domain-specific entity discovery and linking task. Given one review text and one pre-constructed movie knowledge base (MKB) from the douban website, we need to firstly detect all the entity mentions, then link them to MKB's entities. The traditional named entity detection (NED) and entity linking (EL) techniques cannot be applied to domain-specific knowledge base effectively, most of existing techniques just take extracted named entities as the input to the following EL task without considering the interdependency between the NED and EL and how to detect the Fake Named Entities (FNEs) [1]. In this paper, we employ one novel method described in [1] to joint model the 2 procedures as our basic system. Besides it, we also used the basic system's output as features to train models. Finally we ensemble all the models' output to predict FNE. The experiment results show that 80.30% NED F1 score and 93.45% EL accuracy, which is better than that of traditional methods.

Keywords: Fake named entity · Entity linking · Domain-specific knowledge base

1 Task Overview

Named Entity Detection (NED) and Entity Linking (EL) is one key step to bridge unstructured text with structured knowledge base (KB). It is widely studied in this area but mostly for the general KB, and the wikipedia is the most popular study target. Recently domain-specific KB has been found more effective and useful to manage and query knowledge with a specific domain, such as IMDB douban and mtime [1]. The domain specific KB contains more concrete entities. One of the CCKS2016 task is the Domain-Specific Entity Discovery and Linking. It gives one movie knowledge base (MKB) from the douban website, wich contains about 100 thousand star and about 100 thousand movies. The input linking texts are the real people review for people or movies, including short comments, long reviews (more than 1000 characters), topics and synthetic reviews. The training data contains about 870 texts, and the test data contains about 420 texts. Besides this, it also contains 10+ concepts, 30+ properties.

H. Chen et al. (Eds.): CCKS 2016, CCIS 650, pp. 214–218, 2016.
DOI: 10.1007/978-981-10-3168-7_21

2 System Design

Figure 1 is our whole NED and EL system overview for both offline and online process. The system including both offline and online process. The offline is mainly for mining more and more entities' alias to increase the coarse-grained recall. The online process is that given one input text, do the NED and EL steps and get the final result.

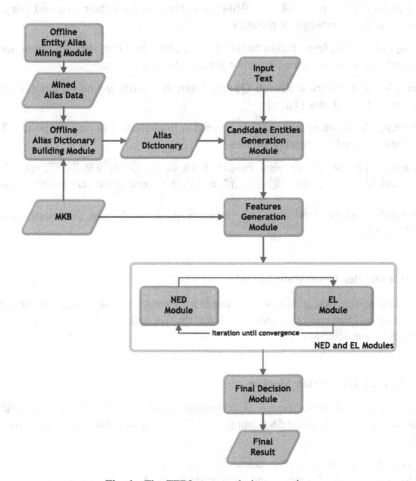

Fig. 1. The TEDL system design overview

2.1 Offline Mining Alias and Dictionary Building Module

The entity alias mining is the key step for the whole system because it directly affect the subsequent modules for its coarse-grained recall. We tried below methods to build our alias dictionary:

Building the Initial Dictionary. From the original MKB we build the initial entity alias dictionary. About 290 thousand entries.

Removing the Noise from Initial Dictionary. There are much noise existing in the initial dictionary, such as "西游记 (新版)", "绝望的主妇 第二季", we should clear them.

Removing Some Very Generic. Alias there are some very generic entity names in the initial dictionary, such as "这个", "时间", we should remove them to avoid bring in much noise in the subsequent modules.

Mining Some Alias from Baidu Baike. From baike's info box, we could mine some alias, and also using the baike's anchor we can also mine some.

Mining Some Alias from Search Query. From the search engine's queries we can mine some entity aliases [7].

Correcting the Spelling Error. This method is implemented during the online. The main idea is the edit distance algorithm.

Generate Alias for the Foreign People. Such as the "尼古拉斯.凯奇", split their names and keep the "凯奇" "尼古拉斯", and then remove some very generic names.

After finishing all the above steps, the final entry number in the alias dictionary is about 460 thousand.

2.2 Candidates Generation Module

After building the alias dictionary, we use it to generate the candidates for one input entity mention. The main data structure is the trie tree and the edit distance algorithm for speller error detection.

2.3 Feature Generation Module

We treat the NED as the binary classification problem and the EL as the ranking problem. So we create about 56 features for NED model and about 17 features for EL model.

EL Model Features. Including below features:
    The popular, WLM, jaccard and content similarity, 5 features in all [1].
    The entity's in-link, out-link, is people or not, 3 features in all.
    Whether the movie's actor, director occur in its context, and whether there is movie occurring among the actor's context, 9 features in all.

NED Model Features. Including below features:
    The link probability, WLM, jaccard and link certainty, 5 features in all [1].
    Mean WLM and jaccard, 6 features in all.
    Some segment feature, such as whether it is one phrase.

CRF features. We trained one CRF model using the training data.

Some context feature. Such as the context mention number, and above EL's c features.

Some mined popular people and video as feature.

2.4 NED and EL Module

For the domain specific KB, the key issue is the Fake Named Entity (FNE) [1], so to overcome this we employ the iteration process describe in [1]. More details about the iterative NED and EL models' training and evaluation in [1].

2.5 Final Decision Module

After the iteration process, we leveraged the boosting idea to train some other models to predict jointly using different training algorithm. So for the EL model, we trained one GBDT classification model and one learning to rank model. One is use the same features as the iterative EL model, and the other uses the features with NED dependent features removed. For the NED models we did the same things, one SVM model and one GBDT model, one EL dependent and one EL independent.

3   Experiments and Evaluation

To assess our system's performance, we build 2 baselines to compare. We refer to our TEDL as Treatment, Baseline1 is make the max iteration number as 1 and removed the final decision module, which is equivalent to the traditional process; baseline2 is the same as the treatment except removing the final decision module, which means that we use the iteration's result as the final result.

Table 1. Font sizes of headings. Table captions should always be positioned *above* the tables.

Approach	NED			EL	Overall(NED + EL)		
	Precision	Recall	F1	Accuracy	Precision	Recall	F1
Baseline1	74.00%	76.41%	75.19%	91.00%	67.34%	69.53%	68.42%
Baseline2	76.85%	79.21%	78.01%	92.41%	71.02%	73.18%	72.08%
Treatment	**79.33%**	**81.30%**	**80.30%**	**93.45%**	**74.13%**	**75.98%**	**75.43%**

Table 1 shows the results. From the table we could see that the treatment has the best performance. Comparing the baseline2 and the baseline 1 we can see that the iteration process achieve +3.66% overall F1 score; comparing the treatment and the baseline2 we can see that adding the +3.35% overall F1 score, and for NED both the precision and recall increased. From the results we could draw the conclusion that the iteration process and the boosting method (final decision module) help a lot. The treatment's result is the final result we submitted.

4 Related Work

The NED and EL problem attracts a lot of people study recent years because it is the key step for many KB applications. The first system to figure out this problem is described by Bunescu and Pasca [2]. The system uses the wikipedia articles as the KB and view all the links as the unambiguous mentions of entity. [3, 4] uses the learning to rank method to perform the EL's candidates ranking, and gets good results. [3] formulates the whole EL process as 4 sub modules: query processing, candidates generation, candidates ranking and top1 candidate validation. Most of existing approaches focus on the general purpose knowledge bases [1]. Many previous systems employed a pipeline frameworks [5, 6]. But in this paper we employed one novel method to model the 2 steps jointly, which is described in [1]. But besides the basic system, we create other subsequent models to predict FNE jointly, achieve good performance.

5 Conclusion

The current traditional EL system focus on the general KB instead of specific domain KB, which has many FNEs. So we employ one novel method which model the NED and EL jointly, which obtain the better result than the traditional methods. We also used the basic system's output as features to train models to predict FNE, the experiment shows that it can achieve better result.

References

1. Zhang, J., Li, J., Li, X.-L., Shi, Y., Li, J., Wang, Z.: Domain-specific entity linking via fake named entity detection. In: Navathe, S.B., Wu, W., Shekhar, S., Du, X., Wang, X., Sean, Xiong, H. (eds.) DASFAA 2016. LNCS, vol. 9642, pp. 101–116. Springer, Heidelberg (2016). doi:10.1007/978-3-319-32025-0_7
2. Bunescu, R., Pasca, M.: Using encyclopedic knowledge for named entity disambiguation. In: Proceedings of the 11th Conference of the European Chapter of the Association for Computational Linguistics (EACL), Trento, Italy, pp. 9–16. Association for Computational Linguistics (2006)
3. Zheng, Z., Li, F., Huang, M., Zhu, X.: Learning to link entities with knowledge base. In: NAACL, pp. 483–491 (2010)
4. Ceccarelli, D., Luccchese, C., Orlando, S., Perego, R., Trani, S.: Learning relatedness measures for entity linking. In: CIKM, pp. 139–148 (2013)
5. Ratinov, L., Roth, D., Downey, D., Anderson, M.: Local and global algorithms for disambiguation to wikipedia. In: HLT 2011, pp. 1375–1384 (2011)
6. Sil, A., Cronin, E., Nie, P., Yang, Y., Popescu, A.M., Yates, A.: Linking named entities to any database. In: EMNLP-CoNLL 2012, pp. 116–127 (2012)
7. Shi, B., Sun, L., Han, X.: Graph based alias extraction using query log. J. Chin. Inf. Process. 27(5) (2013)

Knowledge Base Completion via Rule-Enhanced Relational Learning

Shu Guo[1,2], Boyang Ding[1,2], Quan Wang[1,2(✉)],
Lihong Wang[3], and Bin Wang[1,2]

[1] Institute of Information Engineering, Chinese Academy of Sciences, Beijing, China
{guoshu,dingboyang,wangquan,wangbin}@iie.ac.cn
[2] University of Chinese Academy of Sciences, Beijing, China
[3] National Computer Network Emergency Response Technical Team Coordination
Center of China, Beijing, China
wlh@isc.org.cn

Abstract. Traditional relational learning techniques perform the knowledge base (KB) completion task based solely on observed facts, ignoring rich domain knowledge that could be extremely useful for inference. In this paper, we encode domain knowledge as simple rules, and propose rule-enhanced relational learning for KB completion. The key idea is to use rules to further refine the inference results given by traditional relational learning techniques, and hence improve the inference accuracy of them. Facts inferred in this way will be the most preferred by relational learning, and at the same time comply with all the rules. Experimental results show that by incorporating the domain knowledge, our approach achieve the best overall performance in the CCKS 2016 competition.

Keywords: Knowledge base completion · Relational learning · Rules

1 Introduction

Knowledge bases (KBs) have become extremely useful resources for many NLP related applications like word sense disambiguation [1] and information extraction [2]. They provide large collections of facts about entities and their relations, typically stored as triples, *e.g.*, (Beijing, capitalOf, China). Although such KBs can be very large, they are still quite incomplete, missing large percentages of facts. KB completion, *i.e.*, automatically inferring missing facts from existing ones, has thus attracted increasing attention. Various relational learning techniques have been proposed for this task [3,4,7,9,11,12], significantly improving the ability of KGs in reasoning.

Most existing relational learning techniques, *e.g.*, the embedding-based TransE model [4] and the path ranking algorithm (PRA) [9] make inferences based solely on facts in KBs. They ignore rich domain knowledge which might also be useful for inference. For example, given the fact (Beijing, capitalOf, China), one can easily infer that Beijing cannot be the capital of any country other than China, by using the domain knowledge about capitalOf. Domain

© Springer Nature Singapore Pte Ltd. 2016
H. Chen et al. (Eds.): CCKS 2016, CCIS 650, pp. 219–227, 2016.
DOI: 10.1007/978-981-10-3168-7_22

knowledge is usually encoded as rules, and has been applied in a variety of inference tasks [13–15].

In this paper, we propose rule-enhanced relational learning, specifically rule-enhanced TransE and PRA for KB completion. The key idea is to incorporate additional rules (*i.e.*, domain knowledge) to further refine the inference results given by TransE and PRA, and hence enhance the inference accuracy of them. Facts inferred in this way will be the most preferred by the relational learning techniques, and at the same time comply with all the rules.

2   Related Work

KG completion is to automatically infer missing facts by leveraging existing ones. Early relational learning work on this problem can be roughly divide into three groups: (ii) embedding-based models which encode entities and relations into a latent vector space and model the plausibility of each fact in that space [3,4,6,8]; (2) path ranking algorithms (PRA) which use paths connecting entity pairs to predict potential relations between them [9–11]; (iii) probabilistic graphical models which make inferences via markov logic networks [16] or probabilistic soft logic [17]. This paper focus on first two work. The first one is simple and efficient [4], while the second one is easily interpretable [14].

Most existing relational learning techniques [3,4,8,9] merely focus on facts in KBs, paying less attention on the rich domain knowledge which might also be useful for inference. Domain knowledge has been demonstrated to be extremely useful in knowledge base completion [14,18]. This paper will take into account the additional rules (*i.e.*, domain knowledge) to further refine the inference results given by relational learning techniques.

3   Our Approach

3.1   Overview

To enforce rule-enhanced relational learning, our approach unify relational learning techniques and rules in a framework. In this framework, we perform inferences from existing facts via relational learning techniques (either TransE or PRA), and additional rules will be used to further refine the inference results given by those techniques. Facts inferred in this way will be the most preferred by relational learning, and at the same time comply with all the rules. Figure 1 gives a simple illustration of the framework.

In what follows, we will describe the key components of our approach: (1) relational learning techniques of the TransE model and the path ranking algorithm (PRA); (2) rules imposed to further refine inference results.

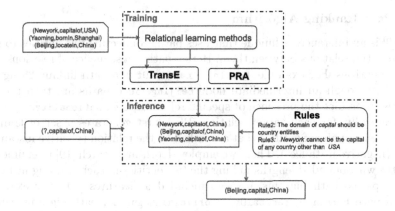

Fig. 1. Simple illustration of Rule-Enhanced Relational Learning.

3.2 TransE Model

TransE [4] is an embedding-based technique which is simple and efficient while achieving state-of-the-art predictive performance. The key idea of TransE is to embed entities and relations in a KB into a continuous vector space, and make inferences in that space.

Specifically, TransE represents entities and relations as low-dimensional vectors in the embedding space. Given a triple (e_i, r_k, e_j) and the embeddings $\mathbf{e}_i, \mathbf{e}_j, \mathbf{r}_k \in \mathbb{R}^d$, TransE assumes that $\mathbf{e}_i + \mathbf{r}_k \approx \mathbf{e}_j$. A score function is further defined on each triple to model its plausibility:

$$f(e_i, r_k, e_j) = -\|\mathbf{e}_i + \mathbf{r}_k - \mathbf{e}_j\|_1. \tag{1}$$

Plausible triples are assumed to have high scores, and low otherwise. To learn these embeddings, a margin-based ranking loss is minimized, *i.e.*,

$$\min_{\{\mathbf{e}\},\{\mathbf{r}\}} \sum_{t^+ \in \mathcal{O}} \sum_{t^- \in \mathcal{N}_{t^+}} \left[\gamma - f(e_i, r_k, e_j) + f(e_i', r_k, e_j') \right]_+ . \tag{2}$$

Here, $t^+ = (e_i, r_k, e_j) \in \mathcal{O}$ is a positive (observed) triple; $\mathcal{N}_{t^+}$ denotes the set of negative triples constructed for t^+, and $t^- = (e_i', r_k, e_j') \in \mathcal{N}_{t^+}$; $\gamma > 0$ is a margin separating positive and negative triples; and $[x]_+ = \max(0, x)$. Stochastic gradient descent (in mini-batch mode) is adopted to solve this problem. In each stochastic iteration, we generate two negative triples for each t^+, one by replacing the head entity and the other the tail entity. To replace a position (head or tail), we use only entities that have appeared in that position (with the same relation). After we obtain the embeddings, for any missing triple, we can predict its plausibility with the score function. Triples with higher scores are more likely to be true.

3.3   Path Ranking Algorithm

PRA [9] is an inference technique that uses paths connecting two entities to predict potential relations between them. Here a path is a sequence of relations that link two entities. For example, `bornIn` $\rightarrow$ `capitalOf` is a path linking `ZhangZiyi` to `China`, through an intermediate node `Beijing`. Such paths are then used as features to predict the presence of specific relations, *e.g.*, `nationality`.

Specifically, for each target relation, PRA first generates a set of training instances, *i.e.*, pairs of entities that are linked by the relation (positive instances) or not (negative instances). Then, we employ depth-first search [19] to enumerate all paths with bounded lengths linking the two entities in each training instance. Besides paths, path bigrams are also included as features [20]. For example, given a path `bornIn` $\rightarrow$ `capitalOf` $\rightarrow$ `officialLanguage` , path bigrams can be generated as `bornIn` $\rightarrow$ `capitalOf`, `capitalOf` $\rightarrow$ `officialLanguage` . The feature values are simply determined by frequency. Finally, we use two-level stacking [21] to combine multiple classifiers, so as to judge whether two entities should be linked by the target relation. We choose 7 base-level classifiers: (1) three decision forest models of random forest [22], ExtraTree [23], and XGBoost [24]; (2) four logistic regression models with different seeds. A meta-level logistic regression classifier is then trained by taking predictions of the base-level classifiers as input features.

3.4   Rules Imposed

We further introduce three types of rules to refine the inference results given by TransE and PRA.

Rule 1 (simple implication). Suppose relation r_1 implicates relation r_2, denoted as $r_1 \mapsto r_2$. Then, any two entities linked by r_1 should also be linked by r_2. For example, `capitalOf` $\mapsto$ `locatedIn`.

Rule 2 (argument type restriction). Arguments of a relation should be entities of certain types. For example, the tail argument of the relation `capitalOf` need to be `Country` entities.

Rule 3 (at-most-one restriction). For 1-To-Many/Many-To-1 relations, the head/tail argument can take at most one entity; for 1-To-1 relations, both arguments can take at most one entity.

By applying these rules directly on observed facts, we obtain additional evidence which can be used to refine the inference results given by TransE and PRA.

4   Experimental Setups

4.1   Data Sets

The released ZHISHI corpus extracted from Zhishi.me [25] consists of three Chinese KBs: BAIDU, HUDONG, and ZHWIKI. Triples in these KBs are not overlapping. For each KB, we split it into two parts, Train-I and Train-II by relation.

Table 1. Statistics of data sets.

Dataset	# Ent.	Train-I		Train-II		# Test-lph	# Test-lpt	# Test-tc
		# Rel.	# Trip.	# Rel.	# Trip.			
BAIDU	86,272	6	40,967	381	566,028	24,613	20,252	75,991
HUDONG	418,529	5	328,927	298	4,679,917	114,928	74,857	176,598
ZHWIKI	144,314	9	17,266	2,819	1,163,405	72,719	86,607	155,772

Train-I contains name-related relations like `chineseName`. Such relations always associate similar entity pairs, *e.g.* (`city-Of-Beijing, chineseName, Beijing`), which can be handled simply by string matching, and hence are not included in relational learning. The other relations are contained in Train-II. We further split Train-II into a training set and a validation set with nearly 5000 triples, used for model training and parameter tuning respectively. Test data is released separately. Table 1 gives some statistics of the data sets, where # Test-lph/# Test-lpt/# Test-tc denotes the number of test triples used for link prediction of head entities, link prediction of tail entities, and triple classification respectively.

For Rule 1, we manually create 5/8/4 simple implication rules for BAIDU, HUDONG, and ZHWIKI (see Table 2 for examples). With these rules, we can perform pure logical inference on existing training triples to infer new facts. For Rule 2, by following the closed-world assumption, we assume that the head/tail argument of a relation can take only entities that have appeared in the same position with that relation. For Rule 3, to identify the relation type (*i.e.*, 1-To-Many, Many-To-1, or 1-To-1), we compute the average number of heads (tails) per tail (head). If the average number is smaller than 2, we label the head (tail) argument as "1" or "Many" otherwise.

Table 2. Examples of rules created.

$\forall x, y$: /baidu/开发商$(x, y) \Rightarrow$ /baidu/发行商(x, y)
$\forall x, y$: /baidu/知名校友$(x, y) \Rightarrow$ /baidu/毕业院校(y, x)
$\forall x, y$: /hudong/常用语言$(x, y) \Rightarrow$ /hudong/官方语言(x, y)
$\forall x, y$: /hudong/籍贯$(x, y) \Rightarrow$ /hudong/出生地(x, y)
$\forall x, y$: /zhwiki/currentClub$(x, y) \Rightarrow$ /zhwiki/Club(x, y)
$\forall x, y$: /zhwiki/上一节目$(x, y) \Rightarrow$ /zhwiki/下一节目(y, x)

4.2 Link Prediction

This task is to complete a triple (e_i, r_k, e_j) with e_i or e_j missing, i.e., predict e_i given (r_k, e_j) or predict e_j given (e_i, r_k). TransE is used for this task.

Evaluation protocol. For each test record $(?, r_k, e_j)$ or $(e_i, r_k, ?)$, we take every entity e' in the dictionary as a candidate answer and calculate its plausibility. If r_k is name-related, the plausibility is determined by the string similarity between

e' and e_j/e_i, defined as $1 - \text{edit}(e',e)/\max(|e'|,|e|)$, where $\text{edit}(\cdot)$ denotes edit distance between string pairs, and $\max(\cdot)$ the longest length among them. If r_k is non-name-related, the plausibility is the score given by TransE according to Eq. 1. Ranking the plausibility in descending order, we get a list of candidate answers. For each candidate answer, if the resultant triple can be directly inferred by Rule 1, we boost it to the top of the list; and if the triple violates Rule 2 or Rule 3, we remove it from the list. We then return the top 200 candidates and record the rank of the correct answer (not released).[1] Aggregated over all test records, we report: (1) the averaged rank (Mean), and (2) the proportion of ranks no larger than n (Hits@n).

Implementation details. We create 100 mini-batches on each KB. The best model is selected by early stopping on validation sets (by monitoring $S = 30\,\% \times \left(1 - \frac{\text{Mean}}{200}\right) + 30\,\% \times \text{Hits@10} + 10\,\% \times \text{Hits@3}$), with a total of at most 1000 iterations. The optimal configurations are: the dimension of the embedding space $d = 70$, the margin $\gamma = 4$, the learning rate for entity $\eta_e = 0.005$, and for relation $\eta_r = 0.0001$ on BAIDU; $d = 70$, $\gamma = 2$, $\eta_e = 0.005$ and $\eta_r = 001$ on HUDONG; $d = 70$, $\gamma = 5$, $\eta_e = 0.001$ and $\eta_r = 0001$ on ZHWIKI.

4.3 Triple Classification

This task is to verify whether a given triple $\langle e_i, r_k, e_j \rangle$ is correct or not. Both TranE and PRA are used for this task.

Evaluation protocol. Given a test triple (e_i, r_k, e_j), we take it as positive if it can be directly inferred by Rule 1, and negative if it violates Rule 2 or Rule 3, without further prediction. For name-related relations, we simply use string matching. A triple is predicted to be positive if the string similarity between the two entities is higher than 0.7. The other relations are handled by either TransE or PRA. For TransE, a triples is predicted to be positive if its score is above a relation-specific threshold δ_r; while for PRA, we can just use the (meta-level) classifier trained for each relation. We choose accuracy (Acc) as the evaluation metric. To enable model selection, for each triple in validation sets, we construct a negative triple by randomly corrupting the entities. Relations with Acc higher than 75\% on validation sets are handled by PRA, and the others by TransE.

Implementation details. For TransE, δ_r is determined by maximizing Acc on validation sets, again, with a total of at most 1000 iterations. The other hyperparameters are set to the optimal configurations as used in link prediction. For PRA, during training, we generate two negative instances for each positive one, one by corrupting the head, and the other the tail. The maximum path length is set to 3, and the path feature values are normalized by Z-score. On BAIDU, we use stacking to train a meta-level classifier. To generate training sets for learning the meta-level classifier, 5-fold cross validation procedure is applied here. The number of trees nt is set to 300 for random forest, 300 for ExtraTree, and 1000 for XGBoost in the base-level classifiers. On HUDONG and

---

[1] If the correct answer is not included in the 200 candidates, we give it a rank of 201.

ZHWIKI, we use standard random forest, with nt set to 1000. All the classifiers are implemented using publicly available tools [2].

5   Results

The experimental results on the three KBs are aggregated and summarized in Table 3. We can see that our approach performs quite well on both tasks, achieving the best overall performance in the CCKS 2016 competition. (The overall performance is evaluated as $30\% \times \left(1 - \frac{Mean}{200}\right) + 30\% \times \text{Hits@10} + 10\% \times \text{Hits@3} + 30\% \times \text{Acc.}$) The results demonstrate the superiority of incorporating domain knowledge into traditional relational learning.

Table 3. Link prediction and triple classification results on the test data of ZHISHI.

Test-lph			Test-lpt			Test-tc	Overall
Mean	Hits@3 (%)	Hits@10 (%)	Mean	Hits@3 (%)	Hits@10 (%)	Acc (%)	
4.15	95.50	96.98	20.30	51.954	68.09	67.69	80.61

6   Conclusion and Future Work

In this paper, we propose rule-enhanced relational learning, specifically rule-enhanced TransE and PRA for KB completion. The key idea of our approach is to use rules as domain knowledge to further refine the inference results given by TransE and PRA. Facts inferred in this manner will be the most preferred by relational learning, and at the same time comply with all the rules. Experiments demonstrate that incorporating domain knowledge into traditional relational learning could really improve the inference accuracy. With our approach, we get the best overall performance in the CCKS 2016 competition on KB completion task. For future work, we would like to try more predictive models for KB completion. For example, combine embedding models and path ranking algorithms via an integrated latent- and path-based features for relation-specific classifier.

Acknowledgements. We are grateful to the many people who made their code available on-line. We also thank the CCKS 2016 organizers for a fun and exciting competition. This research is supported by the National Natural Science Foundation of China (grant No. 61402465) and the Strategic Priority Research Program of the Chinese Academy of Sciences (grant No. XDA06030200).

References

1. Wasserman-Pritsker, E., Cohen, W.W., Minkov, E.: Learning to identify the best contexts for knowledge-based WSD. In: Proceedings of the 2015 Conference on Empirical Methods in Natural Language Processing, pp. 1662–1667 (2015)

---

[2] http://scikit-learn.org/stable/.

2. Hoffmann, R., Zhang, C., Ling, X., Zettlemoyer, L., Weld, D.S.: Knowledgebased weak supervision for information extraction of overlapping relations. In: Proceedings of the 49th Annual Meeting of the Association for Computational Linguistics: Human Language Technologies, pp. 541–550 (2011)
3. Nickel, M., Nickel, V., Kriegel, H.P.: A three-way model for collective learning on multi-relational data. In: Proceedings of the 28th International Conference on Machine Learning, pp. 809–816 (2011)
4. Bordes, A., Usunier, N., GarciaDurán, A., Weston, J., Yakhnenko, O.: Translating embeddings for modeling multirelational data. In: Proceedings of the 27th Annual Conference on Neural Information Processing Systems, pp. 2787–2795 (2013)
5. Wang, Z., Zhang, J., Feng, J., Chen, Z.: Knowledge graph embedding by translating on hyperplanes. In: Proceedings of the 28th AAAI Conference on Artificial Intelligence, pp. 1112–1119 (2014)
6. Lin, Y., Liu, Z., Sun, M., Liu, Y., Zhu, X.: Learning entity and relation embeddings for knowledge graph completion. In: Proceedings of the 29th AAAI Conference on Artificial Intelligence, pp. 2181–2187 (2015)
7. Guo, S., Wang, Q., Wang, B., Wang, L., Guo, L.: Semantically smooth knowledge graph embedding. In: Proceedings of the 53rd Annual Meeting of the Association for Computational Linguistics and the 7th International Joint Conference on Natural Language Processing, pp. 84–94 (2015)
8. Nickel, M., Rosasco, L., Poggio, T.: Holographic embeddings of knowledge graphs. In: Proceedings of the 30th AAAI Conference on Artificial Intelligence, pp. 1955–1961 (2016)
9. Lao, N., Cohen, W.W.: Relational retrieval using a combination of path-constrained random walks. Mach. Learn. **81**(1), 53–67 (2010)
10. Lao, N., Mitchell, T., Cohen, W.W.: Random walk inference and learning in a large scale knowledge base. In: Proceedings of the 2011 Conference on Empirical Methods in Natural Language Processing, pp. 529–539 (2011)
11. Wang, Q., Liu, J., Luo, Y., Wang, B., Lin, C.: Knowledge base completion via coupled path ranking. In: Proceedings of the 54th Annual Meeting of the Association for Computational Linguistics, pp. 1308–1318 (2016)
12. Richardson, M., Domingos, P.: Markov logic networks. Mach. Learn. **62**(1–2), 107–136 (2006)
13. Rocktäschel, T., Singh, S., Riedel, S.: Injecting logical background knowledge into embeddings for relation extraction. In: Proceedings of the 2015 Conference of the North American Chapter of the Association for Computational Linguistics: Human Language Technologies, pp. 1119–1129 (2015)
14. Wang, Q., Wang, B., Guo, L.: Knowledge base completion using embeddings and rules. In: Proceedings of the 24th International Joint Conference on Artificial Intelligence, pp. 1859–1865 (2015)
15. Wei, Z., Zhao, J., Liu, K., Qi, Z., Sun, Z., Tian, G.: Large-scale knowledge base completion: inferring via grounding network sampling over selected instances. In: Proceedings of the 24th ACM International on Conference on Information and Knowledge Management, pp. 1331–1340 (2015)
16. Jiang, S., Lowd, D., Dou, D.: Learning to refine an automatically extracted knowledge base using markov logic. In: Proceedings of the 2012 IEEE International Conference on Data Mining, pp. 912–917 (2012)
17. Pujara, J., Miao, H., Getoor, L., Cohen, W.: Knowledge graph identification. In: Alani, H., Kagal, L., Fokoue, A., Groth, P., Biemann, C., Parreira, J.X., Aroyo, L., Noy, N., Welty, C., Janowicz, K. (eds.) ISWC 2013. LNCS, vol. 8218, pp. 542–557. Springer, Heidelberg (2013). doi:10.1007/978-3-642-41335-3_34

18. Guo, S., Wang, Q., Wang, L., Wang, B., Guo, L.: Jointly embedding knowledge graphs and logical rules. In: Proceedings of the 2016 Conference on Empirical Methods in Natural Language Processing (2016)

19. Shi, B., Weninger, T.: Fact checking in large knowledge graphs: A discriminative predict path mining approach (2015). arXiv:1510.05911

20. Gardner, M., Mitchell, T.: Efficient and expressive knowledge base completion using subgraph feature extraction. In: Proceedings of the 2015 Conference on Empirical Methods in Natural Language Processing, pp. 1488–1498 (2015)

21. Wolpert, D.H.: Stacked generalization. Neural Netw. **5**, 241–259 (1992)

22. Breiman, L.: Random forests. Mach. Learn. **45**(1), 5–32 (2001)

23. Geurts, P., Ernst, D., Wehenkel, L.: Extremely randomized trees. Mach. Learn. **63**(1), 3–42 (2006)

24. Chen, T., He, T.: XGBoost: a scalable tree boosting system. In: Proceedings of the 22nd SIGKDD Conference on Knowledge Discovery and Data Mining (2016)

25. Niu, X., Sun, X., Wang, H., Rong, S., Qi, G., Yu, Y.: Zhishi.Me: Weaving chinese linking open data. In: Proceedings of the 10th International Conference on the Semantic Web, pp. 205–220 (2011)

Knowledge Graph Embedding for Link Prediction and Triplet Classification

E. Shijia[1], Shengbin Jia[1], Yang Xiang[1(✉)], and Zilian Ji[2]

[1] Tongji University, Shanghai 201804, People's Republic of China
e.shijia@gmail.com, {shengbinjia,shxiangyang}@tongji.edu.cn
[2] IBM China Systems and Technology Laboratory,
Shanghai 201203, People's Republic of China
jizilian@cn.ibm.com

Abstract. The link prediction (LP) and triplet classification (TC) are important tasks in the field of knowledge graph mining. However, the traditional link prediction methods of social networks cannot directly apply to knowledge graph data which contains multiple relations. In this paper, we apply the knowledge graph embedding method to solve the specific tasks with Chinese knowledge base *Zhishi.me*. The proposed method has been successfully used in the evaluation task of CCKS2016. Hopefully, it can achieve excellent performance.

Keywords: Knowledge graph · Distributed representation · Entity embedding

1 Introduction

In traditional social networks, the link prediction task is one of the important technologies to discover the relationships among users [1]. Within the link prediction of social network, the *connection* between two users is often said to be a friend relationship. However, in the knowledge graph, the knowledge network is composed of entities and relations. A connection with two entities can be denoted as a triplet (h, r, t), where h is the head entity, t is the tail entity, and the relation between them is represented as r. Different from the social networks, the *connection* in the knowledge graph is usually with a direction, *e.g.* for the triplet *(Yao Ming, born in, Shanghai)*, the relation *born in* is a way from *Yao Ming* to *Shanghai*, but we could not say *Shanghai* was born in *Yao Ming*. Therefore, the traditional link prediction methods used in social networks are not suitable for the link prediction task in the knowledge graph. In addition, because of the flexibility of Chinese language, the rule based natural language precessing (NLP) methods often require a lot of manual intervention.

In this paper, we adopt the representation learning to understand the knowledge graph provided by *zhishi.me*, and embed the entities and relations of the knowledge graph into a low dimensional vector space. The vector representation of the entities and the relations will contain the semantic relationships among them.

H. Chen et al. (Eds.): CCKS 2016, CCIS 650, pp. 228–232, 2016.
DOI: 10.1007/978-981-10-3168-7_23

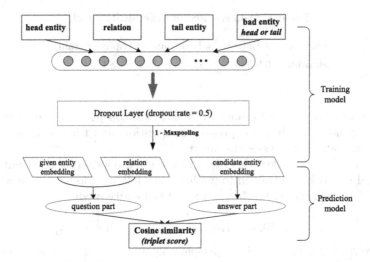

Fig. 1. The overall neural network architecture of our model

The rest of this paper is structured as follows. In Sect. 2, we describe our model architecture used in the evaluation task. In Sect. 3, we summarize the experiment setup of our model. The application of our model is presented in Sect. 4. Section 5 contains related work and finally we give some concluding remarks in Sect. 6.

2 The Embedding Model for Knowledge Graphs

In this section we describe the proposed deep neural networks to solve the LP and TC problems. Figure 1 shows the overall framework of our model. The training part aims to learn the semantic relationships among entities and relations with the negative entities (bad entities), and the goal of the prediction part is giving a *triplet score* with the vector representations of entities and relations. The following is a detailed description.

2.1 Data Preprocessing

The dataset of the evaluation task is from the Chinese knowledge base *zhishi.me*, and the basic statistics of the data are shown in Table 1. In order to meet the requirement of the evaluation task, we first number the entities and relations in turn. During the training time, different IDs represent different entities and relations. This kind of representation can be convenient for us to do the vectorize operations.

2.2 Core Architecture of Knowledge Graph Embedding

For a given triplet (h, r, t) in the training set, our model will learn the vector representations of h and t as well as the r, denoted as **h**, **t** and **r**. The core idea

Table 1. Data set used in the evaluation task

Dataset	#Entities	#Relations	#Triplets (Train)
zhishi.me	644699	3512	7063189

of the model is that transforming the link prediction problem into a question and answer mode, *i.e.* **h** + **r** expresses the question, and **t** is the answer, or **t** - **r** is question, and **h** expresses the answer.

Based on the above ideas, in order to learn the proper vector representations, our neural networks are trained to minimize the following loss function with the training data (illustrated by the example of tail entity prediction):

$$L = max\{0, m - cos(\mathbf{h} + \mathbf{r}, t^{+}) + cos(\mathbf{h} + \mathbf{r}, t^{-})\} \tag{1}$$

where $m > 0$ is the margin hyper-parameter, t^{+} and t^{-} denote the correct tail entity and wrong tail entity respectively. Unlike the TransE [2] or TransM [3] model that use the L_1 or L_2 norm as the dissimilarity measure, we use the cosine similarity (*cos*) to judge the matching degree of *question* and *answer* which can be called as **matching score**. After training with the loss function, it turns out that the loss value of the correct triplet is less than its corresponding wrong triplets. m is used to control the degree of deviation.

During the training process, at every epoch, we randomly sample a wrong entity which is from the whole entity set to each correct triplet in the training set. As a result, the four tuple (h, r, t^{+}, t^{-}) (or (h^{-}, h^{+}, r, t)) forms a training sample. As Fig. 1 shown, we add a *Dropout* layer after the *Embedding* layer to improve the generalization ability of the model and prevent overfitting [4]. Besides that, we add a 1-MaxPooling layer. The vector representation after the pooling layer is treated as the final embedding of the entity or relation which will be used in the loss function.

3 Experiment Setup

In this section, we describe the parameters and experiment environment used in this evaluation task. The parameters need to be fine tuning with different tasks.

3.1 Parameter Settings

In this evaluation task, the margin value m was 0.05, and the embedding dimension of entities and relations was 100. We also tried 200 or 1000 dimensions, and it can get better result on a small dataset (split from training set). However, on the whole dataset, it was more costly. The optimization method employed was Adam [5], and it was more computationally efficient than basic stochastic objective function (SGD). The learning rate was 0.001, and the batch size was 512 per epoch. We trained 200 epochs for the predictions of head entity and tail entity respectively.

3.2 Training Environment

The model used in this evaluation task was implemented with Keras[1]. We used a Tesla K20c GPU device to train the model. Due to time constraints, we believe our model can get better results after longer training time.

4 Applications of the Model

In the triple link prediction tasks, our model would treat all available entities as the candidates for each test sample in the test set $((h, r, \_)$ or $(\_, r, t))$. The trained model would give the matching scores to each *question* and *answer* pairs, and entities ranked at the highest top 200 could be saved as the submitted results.

As to the triplet classification task, we adopted the tail entity prediction model as the test model. For the triplet given by the test set, the model would give the matching score of the test samples. Our strategy was that if the triplet's score was greater than or equal to 0.55, it was considered to be valid, otherwise we tagged it as an invalid one.

5 Related Work

The model used in this evaluation task is related to the following two research areas.

Distributed Representation Learning. It plays an important role with the development of deep learning. The related methods can been applied to various fields, such as NLP, computer vision and image processing [6]. Especially, models based on word embedding have been achieved good performance in the field of text classification [7]. It makes it possible to train on large scale data with limited resources. Inspired by the word embedding model, such as *word2vec*[2], a lot of similar models have emerged recently. *Paragraph Vector* and *Doc2vec* [8] are extensions of *word2vec*, and they learn the vector representations of paragraphs and documents. Essentially, the core of the ideas is to make a good text representation which can express proper semantic information in a specific environment. The embedding models on knowledge graph data also try to catch the key semantic relationships hidden in the numbers of entities, and we can absorb the advantages of those models to help us learn the structure of knowledge graphs.

Knowledge Graph Completion. It aims to predict relations between entities of an existing knowledge graph. It has been several *translation* based methods, such as TransE, TransM, TransR [9] and Hole model [10]. The knowledge graph embedding models with the representative of the TransE have made remarkable achievements in the knowledge graph completion task with the specific datasets.

---

[1] https://keras.io.
[2] https://code.google.com/archive/p/word2vec/.

In essence, all of them try to find out a comprehensive and effective rule which *translates* head entities to tail entities. For the evaluation task in the paper, the scale of the data is far beyond the dataset used in existing experiments. Therefore, we should develop a more effective method to tackle this problem.

6   Conclusion

We describe a deep neural network method with distributed representation to solve the triplet prediction and triplet classification evaluation tasks. Our model can be trained fast with advanced GPU devices and easily extended to other similar tasks.

In addition, the entity candidates in the task is really large. If we can figure out a way to reduce the size of search space, maybe the test result will be better.

Acknowledgement. This work has been partially funded by the National Basic Reseach Program of China (2014CB340404) and the IBM SUR (2015) grant.

References

1. Liben-Nowell, D., Kleinberg, J.: The link-prediction problem for social networks. J. Am. Soc. Inform. Sci. Technol. **58**(7), 1019–1031 (2007)
2. Bordes, A., Usunier, N., Garcia-Duran, A., Weston, J., Yakhnenko, O.: Translating embeddings for modeling multi-relational data. In: Advances in Neural Information Processing Systems, pp. 2787–2795 (2013)
3. Fan, M., Zhou, Q., Chang, E., Zheng, T.F.: Transition-based knowledge graph embedding with relational mapping properties. In: Proceedings of the 28th Pacific Asia Conference on Language, Information, and Computation, pp. 328–337 (2014)
4. Srivastava, N., Hinton, G.E., Krizhevsky, A., Sutskever, I., Salakhutdinov, R.: Dropout: a simple way to prevent neural networks from overfitting. J. Mach. Learn. Res. **15**(1), 1929–1958 (2014)
5. Kingma, D., Ba, J.: Adam: a method for stochastic optimization. arXiv preprint arXiv:1412.6980 (2014)
6. Rumelhart, D.E., Hinton, G.E., Williams, R.J.: Learning representations by back-propagating errors. Cogn. Model. **5**(3), 1 (1988)
7. Kim, Y.: Convolutional neural networks for sentence classification. arXiv preprint arXiv:1408.5882 (2014)
8. Le, Q.V., Mikolov, T.: Distributed representations of sentences and documents. ICML **14**, 1188–1196 (2014)
9. Lin, Y., Liu, Z., Sun, M., Liu, Y., Zhu, X.: Learning entity and relation embeddings for knowledge graph completion. In: AAAI, pp. 2181–2187 (2015)
10. Nickel, M., Rosasco, L., Poggio, T.: Holographic embeddings of knowledge graphs. arXiv preprint arXiv:1510.04935 (2015)

Product Forecasting Based on Average Mutual Information and Knowledge Graph

Zili Zhou$^{(\boxtimes)}$, Zhen Zou, Junyi Liu, and Yun Zhang

School of Physics and Engineering,
Qufu Normal University, Qufu 273165, Shandong, China
zlzhou999@163.com,
996736369@qq.com, 935089344@qq.com, 595037416@qq.com

Abstract. The paper presents a method of modeling the training data which provided by China Conference on Knowledge Graph and Semantic Computing (CCKS) based on average mutual information and knowledge graph. Firstly, calculating the contribution of product attribute to the categories of product, and establishing the product prediction model of product. Then constructing the knowledge graph of training samples which is the network among attributes and categories of product; The average mutual information between attributes and categories is used to provide contribution value for the product prediction model, and the product knowledge graph limits the number of product categories effectively. This is an attempt to integrate algorithm of product forecasting with knowledge graph. After evaluating on the data released by CCKS2016, results show that classification model between average mutual and knowledge graph has high efficiency and accuracy.

Keywords: Conditional probability · Average mutual information · Weight factor · Knowledge graph

1 Introduction

Product forecasting refers to the process of data analysis to get the product category in a given set of data which is not included in the product category, it is a hot spot in big data analysis. Paper [1] is to forecast the carbon emissions from fossil energy consumption, and a model based on artificial neural network is put forward. Aiming at the prediction problem of fast fashion products, a prediction algorithm based on Extreme Learning (ELM) model was constructed in paper [2]. Paper [3] based on ten science and technology factors, to develop and apply the Extreme Learning Machine (ELM) to forecast the gross domestic product (GDP) growth rate. Based on the TRIZ evolution theory, the paper focuses on the technology maturity curve to predict the position of the products on the market at present, which has a significant reference value for the development direction of the product. In [4], Xie et al. gives an improved BASS model which is proposed to be used for short life cycle product demand forecasting. Constantino et al. [5] uses the number of overnight stays in Hotels representative of the tourism demand and artificial neural networks models to forecast it.

© Springer Nature Singapore Pte Ltd. 2016
H. Chen et al. (Eds.): CCKS 2016, CCIS 650, pp. 233–242, 2016.
DOI: 10.1007/978-981-10-3168-7_24

Based on the analysis of literatures and training data released by CCKS2016, the paper produces a classification model based on average mutual information and knowledge graph. Average mutual information provides the contribution of attributes to the categories of product. In another view, the enterprise, origin, custom and destination denotes the concrete entities, so knowledge graph is introduced in our model which can greatly reduce the product categories to be sorted. The method not only can improve the accuracy of category forecast, but also can reduce the amount of computation of the model.

This paper is organized as follows. In Sect. 2, an analysis on the task is discussed, and the three schemes for task solving are presented, the detail description of the schemes is also given in Sect. 3. Schemes are evaluated in Sect. 4, and Sect. 5 is the conclusion of the process of task solving, and the future work is discussed.

2 Task Analysis

There are 18279 training data in CCKS2016 task which have five attributes, that is, Enterprise, Destination, Price, Origin and Custom. The aim of the task is to determine the category of the product data given by CCKS2016. The method which is easily thought and also is our initial scheme is by using conditional probability, however, the scheme has disadvantage which is taking the impacts of attributes on product categories as same. In order to cover the shortage, we improve the initial scheme by using average mutual information, and this is our improved scheme. Furthermore, based on the analysis of attributes, we found that Enterprise, Destination, Origin and Custom, are relative to the entity should rather than pure data. Therefore, the knowledge graph of the product can be constructed, and the knowledge graph of the product can be integrated to the improved scheme, that is our optimal scheme.

3 Task Solution

3.1 Initial Scheme

Conditional probability is the probability that a B event occurs on the basis of the A event, denoted by $p(B/A)$. In order to get the corresponding product category when given an attribute, the scheme calculates conditional probability of product category on the basis of each attribute.

There are 364 product categories in training data. We establish a decision set $A = \{a_1, a_2, \cdots a_{364}\}$, where a_i, $1 \leq i \leq 364$ corresponding to the Product category p001, p002...p364.

3.1.1 Steps of the Scheme

(1) Firstly, the conditional probability of the product category on the basis of enterprise attribute is calculated. There are 560 enterprise codes, that is {1102919313, 1105919182, ..., 440316946}, can be denoted by $E = \{e_1, e_2, \cdots e_{560}\}$, Corresponding conditional probability is $p(a_i/e_j)$, where, $1 \leq i \leq 364$, $1 \leq j \leq 560$.

(2) Simultaneously, The conditional probability of the product category on the basis of the other attributes can be calculated.

Destination: $p(a_i/d_k)$, $1 \leq k \leq 144$;
Origin: $p(a_i/o_m)$, $1 \leq m \leq 131$;
Custom: $p(a_i/c_n)$, $1 \leq n \leq 20$
Price: $p(a_i/p_l)$, $1 \leq l \leq 67$, respectively.

Note: The price data is continuous, so we divide them into groups. In the paper, 67 parts are divided according to the ascending order, Each part can be denoted by $p_1, p_2, \ldots, p_{67}$.

(3) Supposing the effect of Enterprise, Destination, Price, Origin and Custom on classification is the same, that is, the weights are 1.

According to the formulation:

$$p(a_i) = p(a_i/e_j) + p(a_i/d_k) + p(a_i/p_l) + p(a_i/o_m) + p(a_i/c_n) \qquad (1)$$

Corresponding value of $p(a_i)$ can be calculated. Sorting the probability from big to small, and takes the top three product categories as result of product forecasting.

3.1.2 Defect of the Scheme

In this scheme, the correct rate of the result on training data is 19.2%, the result of the lower accuracy is that we consider the weights as 1, that is, considering that each attribute's contribution are the same.

3.2 Scheme Improvement

Without considering the attribute's impact on the results, the prediction results of the initial solution are not ideal. Therefore, based on the original scheme, the average mutual information is introduced to determine the weight of the attributes.

The classification model based on average mutual information considers the influence of the condition attributes (Enterprise,Destination,Price,Origin,Custom) to the decision attribute (Product), and the influence degree of each attribute value is calculated [6].

3.2.1 Steps of the Scheme

(1) According to the formula [7]:

$$I(E; a_i) = \sum_{j=1}^{560} p(e_j a_i) \lg \frac{p(a_i|e_j)}{p(a_i)} \qquad (2)$$

The average amount of mutual information between the Enterprise and each product category can be obtained. Use the same method also can calculate the average amount of mutual information between the Destination, Price, Origin, Customs and each product category. That is, $I(D; a_i)$, $I(P; a_i)$, $I(O; a_i)$, $I(C; a_i)$.

(2) Suppose the product category is a_i, and the weighting factors of the five condition attributes are f_i, g_i, x_i, y_i, z_i. According to the average mutual information between the five attributes and the product category, the attribute weight vector $(f_i, g_i, x_i, y_i, z_i)$ can be calculate by the following expressions.

$$f_i = \frac{I(E; a_i)}{I(E; a_i) + I(D; a_i) + I(P; a_i) + I(O; a_i) + I(C; a_i)} \tag{3}$$

$$g_i = \frac{I(D; a_i)}{I(E; a_i) + I(D; a_i) + I(P; a_i) + I(O; a_i) + I(C; a_i)} \tag{4}$$

$$x_i = \frac{I(P; a_i)}{I(E; a_i) + I(D; a_i) + I(P; a_i) + I(O; a_i) + I(C; a_i)} \tag{5}$$

$$y_i = \frac{I(O; a_i)}{I(E; a_i) + I(D; a_i) + I(P; a_i) + I(O; a_i) + I(C; a_i)} \tag{6}$$

$$z_i = \frac{I(C; a_i)}{I(E; a_i) + I(D; a_i) + I(P; a_i) + I(O; a_i) + I(C; a_i)} \tag{7}$$

According to the formula:

$$p(a_i) = f_i p(a_i|e_j) + g_i p(a_i|d_k) + x_i p(a_i|p_l) + y_i p(a_i|o_m) + z_i p(a_i|c_n) \tag{8}$$

The values of $p(a_i)$ are obtained, and the greater the value, the greater the probability of that the record belonging to the category.

3.2.2   Advantage of the Scheme

The calculation method is simple, and has high efficiency and prediction accuracy is better than the initial scheme. The accuracy rate on training data is 30.1%.

3.3   Optimal Scheme

Knowledge graph [8] is the semantic knowledge base structure, is used to describe the concept of the physical world and their relationship in symbolic form. The basic unit is a triple similar to "entity-relationship-entity".

Based on the analysis of training data, attributes of the records such as enterprise, destination, custom and origin are code numbers of entities which denote the concrete enterprise, country of buyer, etc. These inspire us to construct the network among the entities, so the knowledge graph is the most suitable choice, though the prices are pure data, we divide them into 67 groups and take each group as an entity. So we take the advantages of knowledge graph into account, using the concept of knowledge graph, constructing the reticular structure of entities, especially the relations between the product categories with other attributes [9]. The introduction of the knowledge graph can reduce the range of training samples, and also the amount of computation in the matching process, so the efficiency and accuracy of the forecasting results are greatly improved.

The product knowledge graph model is shown in Fig. 1.

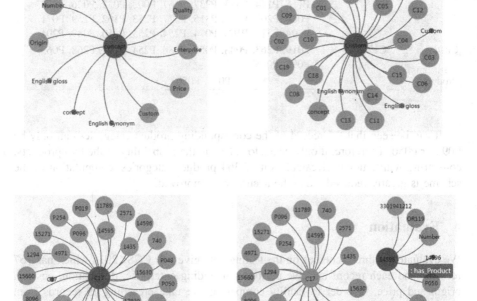

Fig. 1. (a) Concepts of product knowledge graph. (b) Graph of custom attribute. (c) Graph of entity C17. (d) Graph of the 14596th record

For example, take the 14596th records from training data, its attributes are $e_j = 3301941212$, $d_k = D068$, $p_l = 57.7$, $o_m = OR119$, $c_n = C17$, the corresponding product categories related to the each attribute by product knowledge graph are shown in Table 1.

Table 1. Product categories related to the attributes by product knowledge graph

Attributes	Coding	Product categories related to the attributes by product knowledge graph
Enterprise	3301941212	P095, P073, P065, P133, P263, P107, P050, P191, P008, P332
Destination	D068	P050, P101, P095, P077, P335
Price	57.7	P006, P024, P030, P048, P050, P051, P052, P053, P055, P058, P060, P065, P078, P080, P081, P086, P095, P096, P125, P133, P134, P142, P146, P155, P170, P189, P190, P203, P226, P249, P254, P258, P260, P279, P287, P293, P307, P349
Origin	OR119	P161, P187, P047, P082, P073, P086, P234, P095, P119, P228, P291, P092, P018, P357, P031, P033, P114, P120, P164, P140, P180, P126, P115, P110, P150, P286, P263, P068, P262, P292, P048, P235, P045, P356, P065, P354, P116, P144, P343, P237, P167, P008, P032, P026, P173, P064, P203, P349, P315, P327, P253, P162, P139, P071, P304, P319, P108, P074, P254, P255, P250, P089, P201
Custom	C17	P019, P263, P048, P050, P080, P254, P191, P268, P065, P096, P095
Product		P095, P050

It can be seen that the record of the corresponding product categories can only be P095 or P050, Therefore, it only needs to calculate the probability of the two products, comparing with amount of calculation of 364 product categories, computation of the scheme is greatly reduced. Also the accuracy is improved.

4   Evaluation

We evaluate the three schemes on the testing data given by CCSK2016 which has 767 records, and each record has five attributes including enterprise, destination, custom, origin and price. In order to facility comparing, we take preceding 19 records from testing data, compare them by the shooting numbers of product categories.

Table 2 shows the product forecasting result by using initial scheme. From the table, we can see that in product1 column there are 6 product categories are shoot, and 1 product category is shoot in product2 column, 3 product categories are shoot in product3 column. The main reason of low accuracy is that the scheme omitting different attribute has different contribution to the classification.

Table 2. Forecasting result of initial scheme

Number	Enterprise	Destination	Price	Origin	Custom	Product	Product1	Product2	Product3
1	1301930930	D110	4.6	OR028	C18	P185	P173	P226	P185
2	1301965881	D125	10.7274	OR028	C16	P291	P187	P164	P319
3	3104915030	D137	1.8462	OR049	C16	P100	P100	P187	P336
4	3104915030	D140	28.67	OR081	C16	P029	P100	P065	P187
5	3107965839	D043	2.185	OR101	C16	P073	P018	P187	P164
6	3109965344	D056	5.17	OR034	C16	P215	P187	P292	P033
7	3109965344	D082	26.05	OR049	C16	P191	P065	P187	P292
8	3109965531	D137	1.9355	OR094	C16	P082	P082	P187	P217
9	3111966828	D074	7.625	OR091	C16	P351	P351	P098	P187
10	3116961332	D012	5.03	OR057	C16	P068	P006	P068	P187
11	3117940415	D094	3.7697	OR093	C16	P187	P187	P292	P025
12	3117940415	D094	3.4276	OR093	C16	P187	P187	P226	P292
13	3117960753	D113	2.2414	OR091	C16	P187	P187	P032	P076
14	3122262938	D007	30.8	OR001	C01	P332	P025	P047	P048
15	3122266784	D092	4.34	OR100	C16	P292	P025	P187	P065
16	3122268341	D021	1.9741	OR100	C15	P187	P092	P047	P187
17	3122268341	D004	4.53	OR100	C16	P047	P187	P006	P092
18	3122268643	D015	1.9862	OR093	C16	P187	P137	P119	P098
19	3201912922	D033	3.5156	OR049	C16	P292	P226	P228	P187
20	3201912922	D033	3.5156	OR049	C16	P226	P228	P187	P292

The improved scheme takes the contribution value of each attribute into account, and the value is adaptive changing in the process of the forecasting. In Table 3, we can see the number of correct product categories increase from 6 to 8.

Data in Table 4 shows that the number of the correct product categories in product prediction is up to 14, the reason is using product knowledge graph which contains entities of enterprise, origin, custom, etc. and their relations. The use of knowledge graph can remove some product categories which is irrelevant to the attributes of the record, so it can improve the efficiency and accuracy.

Table 3. Product forecasting results of improved scheme

Number	Enterprise	Destination	Price	Origin	Custom	Product	Product1	Product2	Product3
1	1301930930	D110	4.6	OR028	C18	P185	P185	P292	P173
2	1301965881	D125	10.7274	OR028	C16	P291	P164	P187	P319
3	3104915030	D137	1.8462	OR049	C16	P100	P100	P187	P336
4	3104915030	D140	28.67	OR081	C16	P029	P065	P100	P187
5	3107965839	D043	2.185	OR101	C16	P073	P018	P164	P187
6	3109965344	D056	5.17	OR034	C16	P215	P187	P032	P163
7	3109965344	D082	26.05	OR049	C16	P191	P065	P187	P191
8	3109965531	D137	1.9355	OR094	C16	P082	P082	P187	P018
9	3111966828	D074	7.625	OR091	C16	P351	P351	P098	P187
10	3116961332	D012	5.03	OR057	C16	P068	P068	P006	P163
11	3117940415	D094	3.7697	OR093	C16	P187	P187	P025	P027
12	3117940415	D094	3.4276	OR093	C16	P187	P187	P226	P008
13	3117960753	D113	2.2414	OR091	C16	P187	P187	P032	P076
14	3122262938	D007	30.8	OR001	C01	P332	P048	P025	P047
15	3122266784	D092	4.34	OR100	C16	P292	P187	P025	P027
16	3122268341	D021	1.9741	OR100	C15	P187	P092	P187	P047
17	3122268341	D004	4.53	OR100	C16	P047	P187	P006	P092
18	3122268643	D015	1.9862	OR093	C16	P187	P137	P098	P187
19	3201912922	D033	3.5156	OR049	C16	P292	P226	P187	P228

Figure 2 gives the accuracy of the three scheme, it is easy to see by using knowledge graph can get highly improve the accuracy without change the forecasting algorithm in improved scheme. The result proves its advantage in data processing. Knowledge graph is suitable for the data contains entities (no pure data), and the relations can construct based on the given data, on the base of knowledge graph, the data processing algorithm can be optimized and its performance can be improved. The paper is our preliminary attempt, how to mix knowledge graph with data processing rather than integrate them is our future work.

Table 4. Product forecasting results of optimal scheme

Number	Enterprise	Destination	Price	Origin	Custom	Product	Product1	Product2	Product3
1	1301930930	D110	4.6	OR028	C18	P185	P185	P184	P173
2	1301965881	D125	10.7274	OR028	C16	P291	P291	P164	P319
3	3104915030	D137	1.8462	OR049	C16	P100	P100	P187	P234
4	3104915030	D140	28.67	OR081	C16	P029	P100	P065	P336
5	3107965839	D043	2.185	OR101	C16	P073	P018	P292	P234
6	3109965344	D056	5.17	OR034	C16	P215	P215	P033	P292
7	3109965344	D082	26.05	OR049	C16	P191	P191	P351	P065
8	3109965531	D137	1.9355	OR094	C16	P082	P082	P092	P187
9	3111966828	D074	7.625	OR091	C16	P351	P351	P098	P263
10	3116961332	D012	5.03	OR057	C16	P068	P068	P006	P163
11	3117940415	D094	3.7697	OR093	C16	P187	P187	P167	P092
12	3117940415	D094	3.4276	OR093	C16	P187	P187	P167	P092
13	3117960753	D113	2.2414	OR091	C16	P187	P187	P032	P076
14	3122262938	D007	30.8	OR001	C01	P332	P187	P025	P047
15	3122266784	D092	4.34	OR100	C16	P292	P025	P187	P114
16	3122268341	D021	1.9741	OR100	C15	P187	P187	P234	P092
17	3122268341	D004	4.53	OR100	C16	P047	P047	P184	P228
18	3122268643	D015	1.9862	OR093	C16	P187	P187	P234	P137
19	3201912922	D033	3.5156	OR049	C16	P292	P228	P292	P082

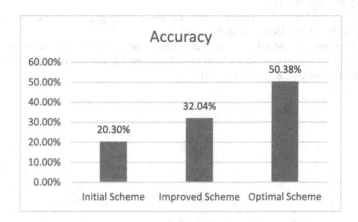

Fig. 2. Accuracy comparison chart of three schemes

5 Conclusion

This paper illustrated the process of how to solve the task which given by CCKS2016. Three schemes were given in the paper, the initial scheme is only using the conditional probability of product category on the basis of other attributes, the result accuracy is low; base on the initial scheme, average mutual information is introduced which is taken as the contribution of attribute to product category, the improved scheme improved the accuracy; The innovation of the paper is that we optimized the improved scheme by using knowledge graph, the motivation is some attribute is not pure data but a mapping of entity in the world, the introduce of the knowledge graph greatly improved the efficiency and accuracy of the scheme.

In future work, we would mix knowledge graph with deep learning or CNN methods etc. in product forecasting, furthermore, we would keep improve the knowledge graph it order to make it be used in data processing easily.

References

1. Wang, Z.X., Ye, D.J.: Forecasting Chinese carbon emissions from fossil energy consumption using non-linear grey multivariable models. J. Clean. Prod. 148 (2016)
2. Wu, J., Zheng, S.: Forecasting of fast fashion products based on extreme learning machine model and Web search data. J. Comput. Appl. **2**, 146–150 (2015)
3. Marković, D., Petković, D., Nikolić, V., et al.: Soft computing prediction of economic growth based in science and technology factors. Phys. A Stat. Mech. Appl. **465**, 217–220 (2017)
4. Jianzhong, X., Yu, Y., Qian, C., Fei, L.: Demand forecasting model for short life cycle products based on improved BASS. Comput. Integr. Manuf. Syst. **21**, 48–56 (2014)
5. Constantino, H.A., Fernandes, P.O., Teixeira, J.P.: Tourism demand modelling and forecasting with artificial neural network models: the Mozambique case study. Tékhne (2016)
6. Ji, C., Hong, T.: New Internet search volume-based weighting method for integrating various environmental impacts. Environ. Impact Assess. Rev. **56**, 128–138 (2016)
7. Zhang, Z., Xuegang, H.: Classification model based on mutual information. J. Comput. Appl. **31**(6), 1678–1680 (2011)
8. Liu, Q., Li, Y., Duan, H., Liu, Y., Qin, Z.: Knowledge graph construction techniques. J. Comput. Res. Dev. **53**(3), 582–600 (2016)
9. Krause, S., Hennig, L., Moro, A., et al.: Sar-graphs: a language resource connecting linguistic knowledge with semantic relations from knowledge graphs. Web Semant. Sci. Serv. Agents World Wide Web **37–38**, 112–131 (2016)

Product Prediction with Deep Neural Networks

E. Shijia and Yang Xiang[✉]

College of Electronics and Information Engineering,
Tongji University, Shanghai 201804, People's Republic of China
e.shijia@gmail.com, shxiangyang@tongji.edu.cn

Abstract. In this paper, we give a solution to the product prediction shared task of CCKS 2016. The main purpose of the task is to determine the product categories for the import and export transaction record data. For this specific dataset, we apply deep neural networks to solve the multi-label classification problem. On the training set, our proposed method achieves a precision of 0.90, and the proposed model can have a good performance on the test set.

Keywords: Multi-label classification · Neural networks · Product prediction

1 Introduction

For the classification problem, traditional methods are focus on learning from a set of examples with only single label, called the binary classification. Nowdays, more classification tasks are often multi-label classification problems. In those tasks, the examples usually belong to more than two categories, even hundreds of categories. In this evaluation task, the training data contains seven basic attributes, of which there are two numeric fields: *Quality* and *Price*, five discrete attributes: *Enterprise, Destination, Origin, Custom* and *Product*. The *Product* field is the target field for the prediction task, and the remainder of the attributes is known to the training attribute. However, the test set of the evaluation task does not contain the attribute of *Quality*. Therefore, for this product prediction task, we have not used the *Quality* attribute as an input feature during the model training process.

In this paper, according the existing data size, we directly use a multi-layer perceptron (MLP) neural network architecture. After 5000 epochs, the accuracy of the training data can reach 90%.

The rest of this paper is structured as follows. In Sect. 2, we describe our model architecture used in the evaluation task. In Sect. 3, we summarize the experiment setup with a discussion of our model. Section 4 contains related work and finally we give some concluding remarks in Sect. 5.

© Springer Nature Singapore Pte Ltd. 2016
H. Chen et al. (Eds.): CCKS 2016, CCIS 650, pp. 243–247, 2016.
DOI: 10.1007/978-981-10-3168-7_25

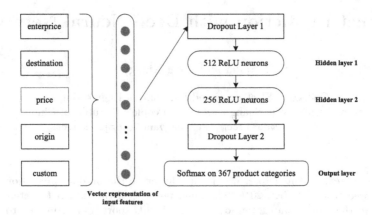

Fig. 1. Architecture of our proposed model

2   The Deep Neural Network Model for Product Prediction

In this section we describe our solution to this specific task. Our main idea is to design an end-to-end method with as little feature engineering as possible, even with no model ensemble. We will give a detail discussion of several models in Sect. 3.2. The final architecture of our model is demonstrated in Fig. 1.

2.1   Data Preprocessing

In order to allow the data to be trained with the deep neural networks, the numeric attribute (*Price*) and discrete attributes need to be unified into vector representations. The preprocessed data will be treated as the input layer of the neural networks.

There are 18279 training samples and 767 test samples in the provided dataset. Due to the discrete attributes of the current dataset only contain 855 values, we directly apply the one-hot encoding method to the input attributes, *i.e.* each discrete attribute can be expressed as a vector with 855 dimensions, and the continuous numerical attribute *Price* is directly used as another dimension of the vector. Therefore, the input features of each sample can be used a vector with 856 dimensions to express.

2.2   Model Description

The core architecture of our model is based on a MLP which is one of the simplest neural network architectures [1]. Specifically, a MLP consists of an input layer, one or more hidden layers, and an output layer. The input layer is always with fixed size representation of input variables, and the hidden layer is used to calculate the intermediate representation of the input variables. Finally, the output layer is used to give the prediction of the output value.

In the MLP architecture, we use the pre-processed data as the input layer of the neural network, then we add the two hidden layers, and the last layer is output layer based on *Softmax* which is a generalization of the logistic function to fit the multi-label classification. In addition, we add two *Dropout* layers [2] before the first hidden layer and the output layer to prevent the model overfitting.

Based on the above ideas, the objective function used in our neural networks is multi-class log loss, also known as the categorical cross-entropy. It is really a common used loss function in the field of multi-label classification. It can be optimized by stochastic gradient descent. The overall model is just like a linear stack of layers, simple but effective.

2.3   The Output of Our Model

There are 364 categories of the output products. As a result, there are 364 neurons in the output layer based on *Softmax* function. The output of that is a probability distribution over the 364 target categories, and the sum of these probabilities is 1. Therefore, for any given sample in the test set, the model is able to return the probability that the sample belongs to any category, and the categories with the top 3 probabilities among the 364 targets are selected as the final prediction.

3   Experiment Results and Discussions

In this section, we describe the parameters and experiment environment used in this evaluation and the final test results. In addition, we give some discussions of the models we have ever tried.

3.1   Parameter Settings

In this evaluation task, we used two hidden layers, the number of neurons in the first hidden layer was 512, and the second was 256. The activation function we used was *ReLU*, and we initialized the network weights with the normal distribution. The optimization method we choose was *Adam* [3], and it was a variant of the typical stochastic gradient descent (SGD). As mentioned before, we added two *Dropout* layers. The dropout rates are 0.25 and 0.5 respectively. The learning rate was set to be 0.001, and the batch size was 128 per epoch. We trained 5000 epochs with a Tesla K20c GPU device. It just took a few minutes to complete the training phase.

3.2   Results

The proposed method with this shared task got the best performance in the CCKS 2016 competition. For one test sample, the model will give three candidate products sorted by descending order according to their correct probabilities. The evaluation results were shown in Table 1.

Table 1. Accuracy of the MLP based model

#Test samples	Hit@1	Hit@2	Hit@3
767	69.36%	79.40%	83.83%

3.3 Discussions

For this specific task, we also tried more sophisticated neural network architectures, such as the embedding model inspired by natural language processing
(NLP) and something relates to long-short term memory networks (LSTM). The
more advanced neural networks didn't get better results than the original MLP.

For the embedding model, we treated the discrete attributes within a sample
as the words in a sentence. We wanted to learn the hidden relationships among
those attributes and hoped that the relationships can reflect some key features
of the product to help the model do the prediction. But the results showed the
semantic relationships among these attributes were not much valuable. Because
the relevance among these attributes was not particularly strong, the embedding model couldn't play its unique role. As for the LSTM models, we tried to
convert the task into a sequence prediction problem, but we didn't make a good
performance with a longer training time. It was because the product category
was not a input sequence item in the provided samples. Therefore, the memory
network couldn't learn a good understanding of the transaction data.

We could figure out that even a simple model can achieve a satisfiable result,
and to solve certain specific problems, complex models are not always necessarily
required.

4   Related Work

The product prediction of the task is just a type of multi-label classification.
There are several related methods in this research area. [4] proposes a system
based on the k-NearestNeighbor (kNN) classifier for multi-label document classification. Its main shortcoming, however, is for real-world use, where the number
of labels of a new document is indeterminate. Liu and Chen [5] have made a
detailed empirical study of different multi-label classification methods on sentiment classification. We can see that the method with best performance is rely
on a high quality sentiment dictionary. It needs more extra resources to do the
multi-label classification.

Besides the traditional methods, the deep neural networks (DNNs) also have
made a good progress in the field of multi-label classification. Ciregan and Meier
et al. apply the DNNs to image classification [6] and traffic sign classification
[7]. [8] uses the deep convolutional neural network (CNN) for fine-grained image
classification. Apart from the image processing area, the DNNs play a import
role in the field of NLP as well. [9,10] use the CNN for sentiment classification.
[11] uses word embeddings for document classification. All these methods show

that the DNNs can make a better performance with large dataset than the traditional rule based methods. The model proposed in this paper is also an effective attempt in the multi-label classification tasks.

5 Conclusion

In this paper, we have introduced a effective deep neural network model to solve the product prediction task. Our model can perform prediction on any import and export transaction records without product categories. The model is able to deal arbitrary size of data. In addition, our results show that we don't have to be obsessed with complex models. In practice, often simple and effective models can also be achieved satisfiable results.

Acknowledgments. This work has been partially funded by the National Basic Reaseach Program of China (2014CB340404) and the IBM SUR (2015) grant.

References

1. Rumelhart, D.E., Hinton, G.E., Williams, R.J.: Learning internal representations by error propagation. Technical report, DTIC Document (1985)
2. Srivastava, N., Hinton, G.E., Krizhevsky, A., Sutskever, I., Salakhutdinov, R.: Dropout: a simple way to prevent neural networks from overfitting. J. Mach. Learn. Res. **15**(1), 1929–1958 (2014)
3. Kingma, D., Ba, J.: Adam: a method for stochastic optimization. arXiv preprint arXiv:1412.6980 (2014)
4. Luo, X., Zincir-Heywood, A.N.: Evaluation of two systems on multi-class multi-label document classification. In: Hacid, M.-S., Murray, N.V., Raś, Z.W., Tsumoto, S. (eds.) ISMIS 2005. LNCS (LNAI), vol. 3488, pp. 161–169. Springer, Heidelberg (2005). doi:10.1007/11425274_17
5. Liu, S.M., Chen, J.H.: A multi-label classification based approach for sentiment classification. Expert Syst. Appl. **42**(3), 1083–1093 (2015)
6. Ciregan, D., Meier, U., Schmidhuber, J.: Multi-column deep neural networks for image classification. In: 2012 IEEE Conference on Computer Vision and Pattern Recognition (CVPR), pp. 3642–3649. IEEE (2012)
7. CireşAn, D., Meier, U., Masci, J., Schmidhuber, J.: Multi-column deep neural network for traffic sign classification. Neural Netw. **32**, 333–338 (2012)
8. Xiao, T., Xu, Y., Yang, K., Zhang, J., Peng, Y., Zhang, Z.: The application of two-level attention models in deep convolutional neural network for fine-grained image classification. In: Proceedings of the IEEE Conference on Computer Vision and Pattern Recognition, pp. 842–850 (2015)
9. Kalchbrenner, N., Grefenstette, E., Blunsom, P.: A convolutional neural network for modelling sentences. arXiv preprint arXiv:1404.2188 (2014)
10. Kim, Y.: Convolutional neural networks for sentence classification. arXiv preprint arXiv:1408.5882 (2014)
11. Kusner, M.J., Sun, Y., Kolkin, N.I., Weinberger, K.Q.: From word embeddings to document distances. In: Proceedings of the 32nd International Conference on Machine Learning (ICML 2015), pp. 957–966 (2015)

Author Index

Printed in the United States
By Bookmasters